CU01433629

Birds of
MONGOLIA

Dedicated to
Prof. Damdin Sumiya, Prof. Osor Shagdarsuren and Prof. Ayurzana Bold,
pioneers of Mongolian ornithology, meticulous scientists,
committed conservationists and honoured teachers.

HELM FIELD GUIDES

Birds of
MONGOLIA

Gombobaatar Sundev and Christopher Leahy

Illustrated by
Adam Bowley, Tony Disley, Carl D'Silva, Kim Franklin,
John Gale, Alan Harris, Ren Hathway, Dave Nurney,
Christopher Schmidt, Brian Small, Jan Wilczur and Tim Worfolk

H E L M
LONDON • OXFORD • NEW YORK • NEW DELHI • SYDNEY

HELM
Bloomsbury Publishing Plc
50 Bedford Square, London, WC1B 3DP, UK
29 Earlsfort Terrace, Dublin 2, Ireland

BLOOMSBURY, HELM and the Helm logo are trademarks of Bloomsbury Publishing Plc

First published in the United Kingdom 2019

Copyright © Gombobaatar Sundev and Christopher Leahy, 2019
Maps © Gombobaatar Sundev and Mongolica Publishing, 2019
Front cover artwork © Tim Worfolk, 2019
Back cover artwork © John Gale and Dave Nurney, 2019

Plate illustrations © 2019: Adam Bowley 69 (part), 80 (part); Tony Disley 64 (part), 65 (part); Carl D'Silva 13 (part),
14 (part), 15 (part); Kim Franklin 22 (part), 26 (part), 27; John Gale 44–47 (part); Alan Harris 4 (part), 9 (part), 10 (part),
11 (part), 13 (part), 14 (part), 15 (part), 16–22 (part), 23 (part), 30–31, 38, 47 (part), 52, 59 (part), 62 (part), 67 (part),
81, 91, 106; Ren Hathway 84–86, 90 (part); Dave Nurney 1–3, 5 (part), 6 (part), 12, 28–29, 32–37, 39–43, 50–51, 53, 54 (part),
55–59 (part), 63–64 (part), 65 (part), 66–67 (part), 68–69 (part), 70–72, 74 (part), 82–83, 87–88 (part), 89–90 (part),
92 (part), 93–94 (part), 95–105, 107–113; Christopher Schmidt 48–49; Brian Small 54 (part), 62 (part), 73–74 (part),
75–80 (part), 88 (part), 92 (part), 94 (part); Jan Wilczur 4 (part), 5 (part), 6 (part), 7–9 (part), 10 (part), 11 (part);
Tim Worfolk 22 (part), 23 (part), 24–26 (part), 60–61, 90 (part)

Topography illustrations pp. 39–40 © Mike Langman, 2019

Vagrant illustrations pp. 268–270 © Carl D'Silva, Alan Harris and Dave Nurney, 2019

Gombobaatar Sundev and Christopher Leahy have asserted their right under the
Copyright, Designs and Patents Act, 1988, to be identified as Authors of this work

All rights reserved. No part of this publication may be reproduced or transmitted in any form or by any means,
electronic or mechanical, including photocopying, recording, or any information storage or retrieval system,
without prior permission in writing from the publishers

Bloomsbury Publishing Plc does not have any control over, or responsibility for, any third-party websites referred
to or in this book. All internet addresses given in this book were correct at the time of going to press. The authors and publisher
regret any inconvenience caused if addresses have changed or sites have ceased to exist, but can accept
no responsibility for any such changes

A catalogue record for this book is available from the British Library

ISBN: PB: 978-0-7136-8704-0;
ePDF: 978-1-4729-4948-6; ePub: 978-1-4729-4949-3.

2 4 6 8 10 9 7 5 3

Designed by Julie Dando, Fluke Art
Printed and bound in India by Replika Press Pvt. Ltd.

To find out more about our authors and books visit www.bloomsbury.com and sign up for our newsletters

CONTENTS

ACKNOWLEDGEMENTS

The production of this book would not have been possible without the support and commitment of family, friends, colleagues and experts on birds.

Our thanks go to Prof. Damdin Sumiya, Prof. Osor Shagdarsuren, Prof. Ayurzana Bold, Dr Shagdarsuren Boldbaatar, Dr Natsogdorj Tseveenmyadag, Axel Bräunlich and Dr Mark Brazil for their recommendations; and also to Dr Sandagdorj Bayarkhuu, Dorj Usukhjargal and the Mongolian Bird Taxonomy and Rarity Committee for updating the list of bird species and providing information on vagrants. Thanks also for professional advice and recommendations from Prof. Michael Stubbe, Dr Wolf D. Busching (Germany), Prof. Reuven Yosef (Israel), Dr Anita Gamauf (Austria), Prof. Per Alström (Sweden), Prof. Woo Shin Lee (South Korea), Dr Kiyoaki Ozaki and Yoshimitsu Shigeta (Japan), and Rishad Naoroji (India), which were invaluable for improving the scientific and conceptual accuracy of the book.

We would like to express our thanks to the authors' families, including Mrs Battsengel Munkhtuya, Kathryn Leahy and children for their encouragement, help and patience while the book was being researched and written.

The original text and early development of this book would have been impossible to start without the financial support of Tony Whitten (World Bank); Prof. Jonathan Baillie, Dr Bhattacharya B. Gitanjali and Ms Eleonor Monks (Zoological Society of London, UK); Dr Richard Reading (former Director of Conservation Biology, Denver Zoological Foundation, USA); Lynette and Ronny Mitchell (UK); Mrs Susan Sloan (USA); Dr Kate and Nigel Barton (Australia); the Mongolian Ornithological Society and Mongolica Publishing.

At Bloomsbury Publishing, we wish to thank Nigel Redman, Jim Martin and Katy Roper, Julie Dando for her considerable design talents, and Tim Harris and Marianne Taylor for their skilful editing and valuable additions to the text.

We thank all the members of the Biology Department at the National University of Mongolia and especially Prof. Mishig Tsogbadrakh, Prof. Nyam-Osor Batkhuu and Prof. Nergui Soninkhishig, head of the Biology Department of the National University of Mongolia and colleagues of the Steppe Forward Programme and Mongolian Ornithological Society for providing equipment and encouragement. We very much appreciate the help of ornithologists and colleagues at the Mongolian Academy of Sciences for their collaboration and professional recommendation, and for compiling information concerning the distribution and status of bird species. We would also like to send our thanks to members of the Mongolian Ornithological Society, including Biraazana Odkhuu, Dorj Usukhjargal, Chuluunbaatar Uuganbayar, Purevdorj Amartuvshin, Bayandonoi Gantulga, Batkhishig Oyuntuya, Dashnyam Bayanmunkh, Jamsran Munkh-Erdene and Bayarmagnai Yumjirmaa for their technical assistance in compiling maps and information on status and conservation of bird species, and providing photographs of subspecies from the Society's database.

The junior author wishes to gratefully acknowledge the support of the Massachusetts Audubon Society, especially Mr James Baird, and presidents Dr Gerard A. Bertrand and Laura Johnson for supporting the extensive field work in Mongolia undertaken through the MAS Natural History Travel Program as well as the time required to prepare the text for publication. He is also grateful to Jalsa Urubshurow, Founder and President of Nomadic Expeditions, for supporting this and other projects intended to preserve and interpret Mongolia's precious natural heritage. And to his many Mongolian friends and colleagues, most especially Badral Yondon, Gereltuv Dashdoorov, Tumendelger Kumbaa, Baatarnyam Navaansharav and Ider Batbayar, with whom he has enjoyed many great adventures in all corners of this extraordinary country, his deepest thanks and affection.

PREFACE

The Mongolian people have a long tradition of resource conservation and the sustainable use of nature's bounty, which was essential to their nomadic herding culture. This tradition is still seen today in Mongolians' careful husbandry of their herds and selective harvesting of native plants to supply all of their basic needs for food, clothing, shelter, transportation and medicine. They were also among the first people in recorded history to practise wildlife conservation. The Italian explorer Marco Polo reported that the Mongolian emperor Khubilai Khan (1215–94) restricted the hunting of birds and mammals over large areas during their breeding seasons and worked to increase wildlife populations through the cultivation of food plants and the provision of natural shelter. Ancient Mongolian falconers may also have been the world's first bird ringers, using silver bands to identify their birds.

A Brief History of Mongolian Ornithology Beginning in the 18th century, as scientific interest in the natural world captured the imagination of monarchs and academics, wide-ranging expeditions were launched from the West to collect and classify hitherto undescribed bird species from the Mongolian steppes, deserts and mountain ranges, and eventually to map their distribution. These included explorations in 1772–73 by the great German naturalist Peter Simon Pallas (1741–1811), sponsored by Catherine the Great, resulting in the discovery of hundreds of species of plants and animals, some of which still bear his name, including Pallas's Gull, Pallas's Sandgrouse and Pallas's Warbler. Other European – and, eventually, American – travellers and ornithologists visited Mongolia for various periods and with various results, as follows:

Gustav F. R. Radde, Prussian naturalist, visited in 1859 and 1863; Radde's Warbler.

General Nicolai Przewalskii, Russian Cossack naturalist and explorer, 1880 and 1883.

Grigory Potanin, Russian ethnographer and natural historian, 1876–99.

Vladislav S. Molleson, Scottish/Russian explorer-naturalist, 1877–99; listed 206 species from Mongolia.

G.E. Grumm-Grzimailo, 1903; listed 291 species.

V.C. Dorogostoisky, Russian ornithologist, 1907; listed 147 species in Tes and Hövsgöl areas.

Petr P. Sushkin, Russian ornithologist, 1914; listed 219 species.

Arkady Y. Tugarinov, Russian ornithologist, 1915; listed 136 species in Tagna range, Uvs and Tes; and 1929; listed 309 species.

Roy Chapman Andrews, American explorer, adventurer and naturalist, 1919, 1922, 1923 and 1925; collected 79 species.

Elizabeth V. Kozlova, Russian ornithologist, 1923–29; collected 1,700 specimens of 305 species and subspecies.

Charles Vaurie, American ornithologist, 1964; listed 355 species.

As can be seen from the above list, explorers and bird experts from Russia and other foreign countries dominated Mongolian ornithological history until the mid-20th century. However, after the National University of Mongolia was established in 1942, pioneering Mongolian ornithologists, especially Professors Dondog Tsevegmid, Anudari Dashdorj, Osor Shagdarsuren, Ayurzana Bold, Damdin Sumiya, Shagdarsuren Boldbaatar and Natsagdorj Tseveenmyadag conducted extensive field surveys on distribution, occurrence, abundance and bird ecology, collaborating with scientists from Russia, Germany and other countries. They published a number of significant works on Mongolian birds, documenting a total of 457 species belonging to 200 genera, 60 families and 19 orders, including 140 species newly discovered in the country during the previous four decades.

During the period of influence by the Soviet Union (1921–90), the Mongolian ornithologists cited above trained and mentored many younger ornithologists in the country, and since 1998 the National University of Mongolia team led by the senior author has continued to broaden the scope and improve the quality of Mongolian ornithology. Many undergraduate and graduate students under the tutelage of this team have devoted themselves to bird research and bird conservation activities both in Mongolia and abroad, and have published a number of modest bird guides (see References and additional reading).

In 2009, ornithologists from Mongolia and many other countries assessed the entire known Mongolian avifauna, then totalling 487 species, using the criteria of the International Union for the Conservation of Nature (IUCN). The resulting Regional Red List for Mongolian Birds and Summary Conservation Action Plan on Mongolian Birds was published in 2011 – the first comprehensive scientifically-based work on Mongolian birds to be published in English. This project was initiated and supported by the Zoological Society of London, the Mongolian Ornithological Society, and the National University of Mongolia, with involvement from the Mongolian Academy of Sciences and many NGOs. Needless to say, this seminal work, together with the more than two centuries of field studies and reports that preceded it, have been critical resources in the preparation of the present field guide.

Gaps in Our Knowledge and a Plea for Additional Records In preparing this guide the authors have thoroughly reviewed historical and recent ornithological records in Russian, German, English and Mongolian publications, and reached out to foreign birding groups known to have visited the country. Despite these efforts, the reader will note some gaps in the data pertaining especially to vagrants; most of these involve records that the authors believe to be correct, but which lack complete dates and/or locality information, due to confusion about Mongolian place names or simply inadequate record keeping. Since the establishment of the Mongolian Bird Taxonomy and Rarity Committee (2011), all records of vagrants and other rarities known to us have been reviewed and protocols established for submitting records for validation. However, Mongolia is a vast country with limited access to many rich birding areas and still relatively few birders, and we therefore strongly encourage visiting birders to send their records – ideally with photos and/or field notes documenting rarities – to the Mongolian Bird Taxonomy and Rarity Committee, Mongolian Ornithological Society: info@mos.mn, gomboo@num.edu.mn.

This guide documents 417 regularly occurring species, plus 86 vagrants, giving a total recorded avifauna to date of 503 species. There can be no question that with more fieldwork by Mongolians and birdwatchers from elsewhere in the world, this total will increase. It is our hope that this new guide will encourage more interest in and knowledge of Mongolian birdlife – both of which are essential for its conservation.

INTRODUCTION

Mongolia has been called the Last Wilderness Nation. Due to circumstances of geography, culture and history, Mongolia today boasts a combination of spectacular wilderness landscapes, biological diversity and rare species unmatched in the temperate regions of the planet. Lying at the heart of the great northern Asian steppe, it bridges the vast Siberian taiga forests of the north and the world's coldest deserts to the south, and encompasses great mountain ranges, pristine rivers and lakes, as well as its storied plains and arid lands. Though it is a large country with an area about half the size of Western Europe, or slightly smaller than the state of Alaska, it is also one of the most sparsely populated, leaving ample room for wildlife. Its nomadic herding culture is imbued with a love of the land and an ethic that has traditionally valued stewardship of resources over exploitation. Mongolians by and large are shepherds, not hunters. And while the country's encirclement by China and the former Soviet Union has made for many hardships through the centuries, this has also left Mongolia relatively free of the industrial development that has greatly scarred and poisoned the environments of her neighbours.

For all of these reasons it is not unreasonable to think of Mongolia as one of the world's great bird sanctuaries, harbouring an array of globally rare and local species such as Altai Snowcock, Relict Gull, Henderson's Ground Jay and Kozlov's Accentor; supporting diverse and thriving populations of raptors, waterfowl and open-country birds; and providing such moving avian spectacles as swarms of Pallas's Sandgrouse arriving near dawn at desert oases or migratory flocks of Demoiselle Cranes moving gracefully across the steppe. And yet, for most prospective visitors – their imaginations understandably focused on Genghis Khan and camel caravans – Mongolia's wealth of birdlife and other natural wonders remains a 'best-kept secret'.

With the recent discovery of immense deposits of precious minerals, Mongolia faces a future of enormous opportunity coupled with significant risk. Will its new-found wealth be used to ensure the preservation of its precious natural and cultural heritage or finance their destruction? We hope that this first field guide to the birds of Mongolia will give birdwatchers, as well as all lovers of wilderness and the natural world, a sense of how much Mongolia has to offer, and – if its natural heritage is not preserved, what it stands to lose.

MONGOLIA AND ITS BIRDLIFE

GEOGRAPHY

Mongolia – known historically as Outer Mongolia (as distinct from the Chinese province of Inner Mongolia) and during the period of Soviet influence (1921–90) as the Mongolian People's Republic – is a large, landlocked country in northern Central Asia. Bounded by Russia (eastern Siberia) to the north and China to the south, it spans about 2,400 km from east to west and about 1,250 km from north to south at its widest points. It covers an area of 1,564,116 km^2 – about half the size of Western Europe and slightly smaller than Alaska. The country is divided into 21 provinces (aimags), each with a provincial capital. The provinces are, in turn, divided into 329 districts (soums).

Provincial capitals of Mongolia

DEMOGRAPHICS

As of August 2016 the population of Mongolia stood at 3,101,563 people, with 45% living in the capital city of Ulaanbaatar. It has the lowest population density of any nation on Earth (0.5 person/km^2), with potentially positive implications for bird and habitat conservation. Ethnically, the population is predominately Mongol, though there is a significant Kazakh population in the Altai Mountains of the west. The majority of Mongolians are Buddhist, and 30% of the population continues to follow the age-old nomadic herding culture.

TOPOGRAPHY

The country is mountainous, with about 60% of the surface lying between 1,000–2,000 m above sea level, and just 15% below 1,000 m. Mongolia's highest peak reaches 4,374 m, and the country is one of the highest in the world overall because it sits on a high plateau with an average elevation of around 1,500 m.

There are five major mountain ranges in Mongolia. From the highest overall to the lowest, they are: the Mongol-Altai range in the west (highest peak, 4,374 m); the Gobi-Altai range, a continuation of the Mongol-Altai, south and east into the Gobi Desert (4,090 m); the Khangai range in the centre-west (4,008 m); the Hövsgöl (Sayan) range in the north (3,491 m); and the Hentii range in the north-east (2,799 m).

CLIMATE

In addition to being adapted to high and moderate altitudes, Mongolian birds must also survive extremes of cold, heat and aridity. The average January temperature is -20° to -30°C. Even in the region of Lake Uvs in the Great Lakes Depression, some areas of the Khangai mountains and the Gobi, temperatures can fall

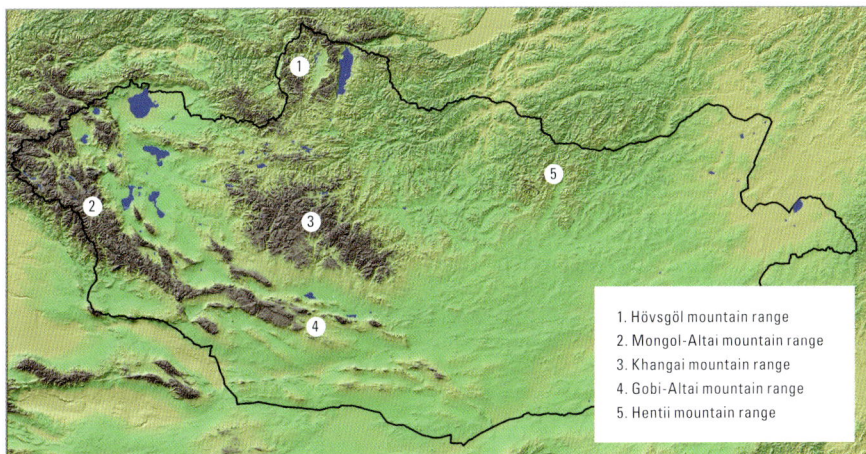

Topographical map of Mongolia

1. Hövsgöl mountain range
2. Mongol-Altai mountain range
3. Khangai mountain range
4. Gobi-Altai mountain range
5. Hentii mountain range

to -40° or -50°C. Lake Hövsgöl in the far north typically remains at least partially frozen until mid-June. While summer temperatures in the north hover around 22°–30°C, the Gobi in summer can reach 40°–50°C. Mongolians proudly refer to their country as the 'Land of Eternal Blue Skies', and it is true that there are as many as 260 sunny days a year. However, since Mongolia has an 'extreme continental climate', which means that it lacks the ocean's moderating influence, the weather can change frequently and often dramatically, with rapid temperature changes as wide as 35°C or a morning blizzard followed by an afternoon warm spell.

Wind is also a characteristic of the changeable Mongolian climate, especially in the Gobi, where fierce dust storms in the spring and autumn may be driven by winds as strong as 140 km/h.

The factor most influential in shaping and colouring the ecological 'face' of Mongolia at all latitudes and altitudes is the relative dearth of precipitation. In some years, no rain at all falls in the driest parts of the Gobi, where recent annual averages of 42 mm have been recorded, and even those northern slopes and valleys that catch moisture-laden winds from the Arctic capture only 200–350 mm of rain per year – one-third to one-quarter the amount that falls on average in northern Europe or north-eastern North America.

ECOLOGICAL ZONES AND HABITATS

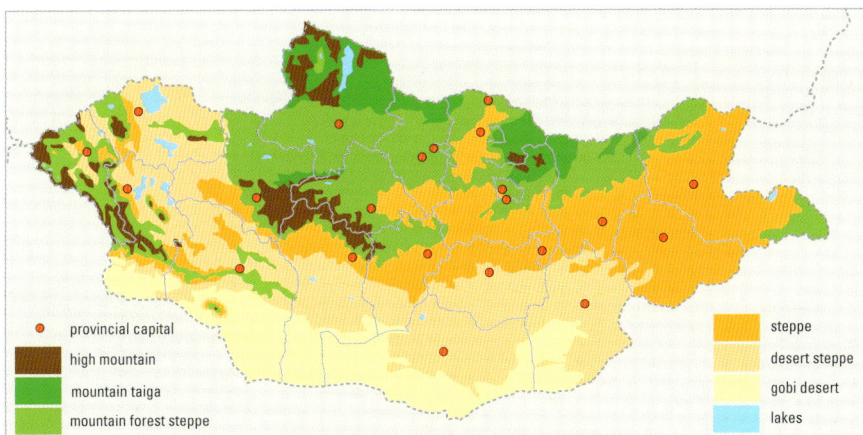

- provincial capital
- high mountain
- mountain taiga
- mountain forest steppe
- steppe
- desert steppe
- gobi desert
- lakes

Ecological zones of Mongolia based on vegetation type and topography

An easy way to visualise the ecological zones of Mongolia is to picture a cloud of (scanty) moisture arriving at the northern border of the country from the north and gradually diminishing as it moves south, until beyond the mountain ranges of the Gobi it has completely evaporated. Thus a traveller moving north to south across the country would begin in relatively lush conifer forest, pass through ever-drier grasslands and end in sandy or stony desert. This tidy pattern is interrupted by several major mountain ranges, which both capture and block moisture regionally. While it may be useful to understand this relationship between Mongolia's climate and its natural communities, it should be borne in mind that the boundaries of the zones defined by scientists are 'soft' and that many bird species, having never heard of them, are quite at home in two or three, or even all of them.

The tundra-like habitat in the northern alpine zone is dominated by sedges, grasses and scattered ericaceous shrubs (*G. S.*)

Alpine zone

In Mongolia, the high treeless zone below the permanent glaciers takes two forms. In the north, in the Altai, Khangai, Hentii and Hövsgöl ranges, stony barrens with matted sedges and grasses and scattered heathy shrubs resemble Arctic tundra and, indeed, share many plant species with the frozen lands at lower altitudes further north. The arid peaks in the Gobi-Altai are even more barren and typically dotted with juniper 'trees', which due to the harsh climate grow outward into broad mats rather than upwards. This high mountain zone, comprising about 3.6% of Mongolia's surface, contains some of the country's most spectacular birds, such as Altai Snowcock, Bearded Vulture, Wallcreeper and Great Rosefinch, as well as Chukar Partridge, Hill Pigeon, Red-billed Chough, Eurasian Crag Martin, Black Redstart, Common Rock Thrush, Pied Wheatear, Alpine, Altai and Brown Accentors, Twite, Mongolian Finch, White-winged Snowfinch, Himalayan Beautiful Rosefinch and Grey-necked Bunting.

Taiga

This refers to the northern coniferous forest, also known as Siberian boreal forest. It is the most extensive forest type in the world, circling the globe in a wide band across Canada, Scandinavia and Russia. It reaches the southern limit of its range in northernmost Mongolia, thriving only in the coldest, wettest part of the country and covering only about 4.5% of its area. It contains variable mixtures of cone-bearing trees plus birch and poplar, but in Mongolia it is often dominated by Siberian Larch, a deciduous conifer that turns the hillsides to gold in autumn. Typical taiga birds include Black-billed Capercaillie, Black (and other)

The desert alpine zone in the Gobi-Altai is characterised by rocky barrens with steep cliffs and broad mats of juniper (*G. S.*)

Woodpeckers, Spotted Nutcracker, several species of leaf warblers, Oriental Cuckoo, Red Crossbill and Pine Bunting. In this and the following zone there are also many species that nest in shrubby copses and other edge habitats rather than in forest, e.g. Siberian Rubythroat.

The taiga is also home to the Tsaatan people of Hövsgöl province, who live in tepees and move camps seasonally with their (native) reindeer herds, like the Sámi in the taiga of northern Scandinavia. They share Turkic lineage with the natives of Tuva in neighbouring Russia.

Larches and other conifers dominate the boreal landscape up to the tree-line in the Siberian taiga, or Siberian boreal forest, in northernmost Mongolia (*C. L.*)

In the northern hills, forest can only thrive on north-facing slopes, which trap moisture, leaving the south-facing slopes in a rain shadow where only grasses survive (*C. L.*)

Forest steppe

South of the taiga and on the lower mountain slopes is a transition zone where forest meets grassland. The hillsides in this zone, which covers much of north-central Mongolia and about 15% of the country as a whole, look as though they have been selectively cut or burned, with forest covering one side and grassy pastureland on the rest. But the effect is actually a natural result of the rainfall pattern. The north-facing slopes catch what little moisture is borne on the prevailing northerly winds and are therefore able to support trees, while the other slopes lie in a rain shadow, where only grasses and herbaceous plants can thrive. The combination makes for a high diversity of wildlife because it attracts open-country species such as Demoiselle Crane, tree-nesting species such as Daurian Redstart and birds such as Booted Eagle and Daurian Jackdaw that use both types of habitat. The relatively moist soils of the forest steppe also make for stunning displays of primroses, pasque flowers, cinquefoils and many other species of wildflowers in late spring and early summer. This zone holds the greatest percentage of Mongolia's human population.

Deciduous woodland occurs as gallery forest along watercourses where there is sufficient moisture and rich, deep soils to support it (*G. S.*)

The eastern steppe of Dornod province is characterised by taller, lusher grasslands than elsewhere in Mongolia (*C. L.*)

Steppe and desert steppe

'Steppe', a term derived from the Ukrainian word for grasslands, occurs naturally where a variety of limiting factors such as low rainfall, poor soils, grazing and/or fire inhibit the growth of trees, yet the biome gets enough rain to support grasses and a few moisture-conserving shrubs. Steppe can be relatively lush and occur in rolling oceanic plains of long grasses, as in Mongolia's eastern steppe where vast herds of Mongolian Gazelles thrive in the largest unfenced grassland in the world, and the increasingly rare Great Bustard can still be seen. In contrast, the arid steppe of the Gobi, with much shorter, sparser grasses, is just a few millimetres of rain short of true desert and has been defined as a separate zone: desert steppe. Between them these two types of steppe make up over 34% of the Mongolian landscape. Many bird species occur commonly in both dry and more humid steppe. Upland Buzzard, Steppe Eagle, Saker Falcon, Demoiselle Crane, Mongolian Lark, Richard's Pipit and Northern Wheatear are typical birds of the lusher steppe, while Pallas's Sandgrouse, Oriental Plover, Isabelline Wheatear and Père David's Snowfinch tend to be more common in the drier version.

Desert steppe is intermediate between lusher steppe, which receives more rainfall, and true desert with less (*C. L.*)

The grazed grasslands of the central Mongolian steppe have been home to nomadic herders for centuries, resulting in a habitat 'managed' by herds of domestic grazing animals (*G. S.*)

Khongoryn Els is one of the world's most spectacular dune systems – 100 km long, 12 km wide and with a maximum height of about 300 m – but it attracts few birds (*C. L.*)

Gobi Desert

When Mongolians talk of the 'Gobi' they are referring to a specific vast dry region (Gobi means 'waterless place') that includes arid steppe as well as true desert and which occupies the southern third of their country and extends southward into neighbouring China. However, the Gobi Desert can also be defined in several other ways. For example, there are a number of geographical regions of the Gobi, e.g. the Dzungarian Gobi of the extreme south-west, the Gobi-Altai mountain range and the Trans-Altai Gobi, terms that are sometimes useful in describing bird distribution.

There are also several different physiographic Gobis. The word 'desert' inevitably evokes visions of sand-dunes, and the Gobi has some magnificent ones, the grandest reaching almost 100 km in length and topping 300 m in height, with 'singing sands' as a special bonus. However, most of the desert's surface is hard-pan or gravel with much exposed bedrock at the surface, often in striking formations sculpted by harsh winds.

The red sandstone formation known as the Flaming Cliffs has yielded an astonishing diversity of dinosaur and other fossils since the Roy Chapman Andrews expedition unearthed the first known dinosaur eggs there in the 1920s (*C. L.*)

Saxaul forest is dominated by a small, salt-tolerant tree *Haloxylon ammodendron*, which occurs from the Middle East to Central Asia (*G. S.*)

Then there is the famously harsh Gobi climate. Despite occasional summer temperatures reaching 50°C, the Gobi must be described overall as a cold desert, the northernmost of its kind in the world, with winter temperatures dropping to -40°C and often accompanied by fierce winds blowing from Siberia. It is also a high desert, its floor lying at 914–1,524 m and the peaks of the Gobi-Altai Mountains ascending to 2,400 m. The Gobi's aridity results from the 'rain shadow' cast by the great wall of the Himalayas to the south, which captures almost all the moisture blown north from the Indian Ocean in the summer monsoon. While the driest parts of the Gobi may have no rain at all in a given year, the region as a whole gets an average of 195 mm, enough to recharge extensive ground water reservoirs. In addition, several large rivers from the Khangai Mountains to the north flow down to the Gobi, ending in broad, shallow, saline lakes.

In ecological terms, 'desert' indicates a near-absence of vegetation, including even the hardiest grasses, which mark the distinction between arid steppe and true desert. The plants that signify true desert have evolved strategies to cope with extreme aridity, such water-retaining (e.g. succulent or fuzzy) leaves and stems, long roots that tap into underground water, salt tolerance or the ability to parasitise the roots of other desert plants for nutrients rather than depending on leaves for photosynthesis. The most prominent indicator plant of true desert is Saxaul (*Haloxylon ammodendron* – literally 'salt wood sand tree'), a small tree that occurs in scattered groves throughout the Gobi. This true desert occupies about 19% of Mongolia.

Most of the common birds of the Gobi occur in arid steppe as well as true desert. However, some of the species that can be broadly thought of as desert birds are Pallas's Sandgrouse, Henderson's Ground Jay, Asian Desert Warbler, Desert Wheatear, Saxaul Sparrow and Mongolian Finch.

Most of the Gobi Desert is covered in small, dense, widely spaced shrubs, whose species mix varies from place to place, depending upon soil type and water availability (*C. L.*)

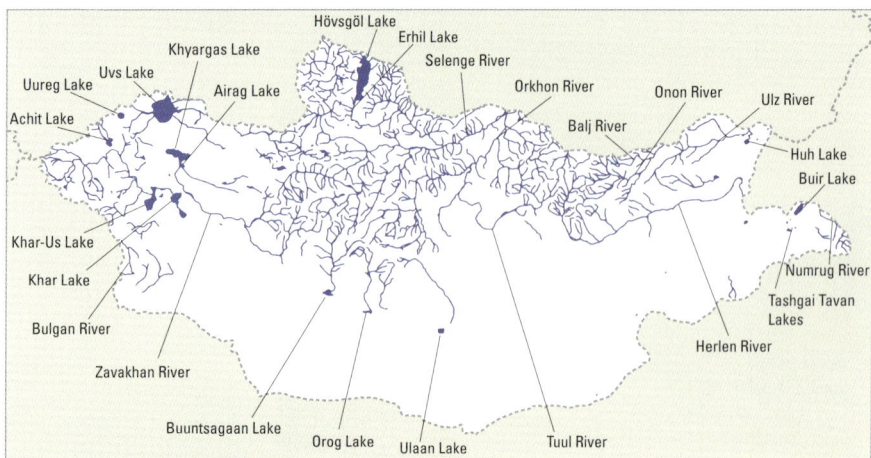

Mongolia's lakes and river systems

Water bodies and wetlands

Despite its low rainfall, Mongolia is rich in water resources, with approximately 1.5 million ha of standing water bodies – both freshwater and saline – and 50,000 km of rivers, not to mention a variety of wetland types such as bogs, wet meadows and reedbeds. Together, these sources of water in a generally dry country attract a wealth of birdlife from the Black-throated Divers and White-tailed Eagles attracted to the deep, fish-filled waters of Lake Hövsgöl, and the hosts of migratory waders and other birds that stop to rest and

The Herlen River is a significant source of water for many migrants. It is one of the largest rivers in Mongolia, running from Siberian taiga forest through open steppe to China (*G. S.*)

Lake Hövsgöl, the largest body of freshwater in Mongolia, is an ancient, nearly pristine rift lake 267 m deep, surrounded by mountain ranges mostly covered with mature larch forest (*G. S.*)

feed around the Gobi's broad, shallow lakes, to rarities such as White-headed Duck, Dalmatian Pelican and Relict Gull, which can be found in remote water bodies, and a rich diversity of parrotbills, reed warblers, buntings and other wetland songbirds in extensive marshes. The most important of these water bodies and wetlands, along with their birdlife, are described below under Birdwatching in Mongolia. It is worth noting that 70% of Mongolia's threatened birds are aquatic and wetland species.

Above: saline lakes occur where rivers originating in the mountains end in closed basins rather than emptying into an ocean (G. S.)

Freshwater lagoons supporting emergent vegetation fringe much of Lake Hövsgöl's shoreline and attract waterfowl, waders and wagtails, among other bird species (C. L.)

Reedbeds and other freshwater marshlands are common in lake basins throughout Mongolia, providing habitat for reed specialists and many waterbirds (G. S.)

BIRD MIGRATION IN MONGOLIA

Only about 11.4% of Mongolia's avifauna consists of permanent residents, that is, species that both nest and winter in the country. The others are either breeding visitors, which nest in Mongolia but migrate south, typically to wintering grounds in southern or coastal Asia, passage migrants, which pass through Mongolia between breeding grounds further north and winter quarters further south, and winter visitors, Arctic or nomadic species that arrive in Mongolia in late autumn and return north in early spring. Thus migration is an enormously significant factor in the lives of 'Mongolian' birds.

The most widely accepted theory on the origins of long-distance bird migration is related to climate change over great periods of time, specifically the advance and retreat of glaciers during periodic ice ages and the warmer intervals between them. Simply stated, the theory holds that as the ice advanced, bird species of the Arctic and temperate zones were forced southwards into areas that today are in the tropics, reinvading their former ranges as the ice retreated but, especially in the case of insectivorous and frugivorous species, continuing to fly south for the winter. A companion theory holds that a few species that evolved in the tropics also moved northward as the climate warmed but returned to their places of origin for the winter.

Other forms of bird movements include seasonal dispersals based on the availability of food supplies, e.g. conifer seed-eaters such as Red Crossbill leaving areas with low productivity to seek areas with larger cone crops, or northern owl populations moving south in response to fluctuating rodent populations. Once these semi-nomadic birds arrive in Mongolia they will also move in response to atmospheric phenomena such as snow cover. In addition, young birds of many species will disperse in all directions at least initially rather than taking direct routes to their wintering grounds. Finally, small numbers of birds of different species end up in Mongolia outside their normal ranges due to climatic effects such as storms or changing environmental conditions. While these species are initially classified as 'vagrants' (see listing on pp. 268–270), some may turn out to have been 'pioneers', extending their species' range into new territories.

Different types of birds use different strategies to undertake long-distance migrations. Most waders and songbirds, for example, travel mainly at night at elevations as high as 1,500–9,000 m, finding their way by using star patterns and signals from Earth's magnetic field. These nocturnal migrants, finding themselves over the Gobi Desert at dawn, often descend and become concentrated in the limited patches of vegetation or at the few scattered oases. Or they may be forced down by a sudden spring blizzard and land in a 'fall' involving thousands of birds. Both of these scenarios can result in very exciting birdwatching. Migratory soaring birds such as eagles and storks migrate by day, using a different technique; buoyed by columns of warm air called thermals generated from the sun-warmed earth, they rise aloft with minimal energy and then drift off in the direction of their migratory path, finding another thermal when they lose altitude, an efficient strategy that minimises energy consumption. These soaring migrants also take advantage of updrafts of air along mountain ridges and follow 'leading lines', visual clues such as river valleys and lake shores, and can often be seen in numbers by birdwatchers stationed along these flyways.

While we understand basic patterns of bird migration, the exact routes and destinations of migrants have remained obscure. However, recent technological advances have facilitated the study of bird migration in Mongolia in recent years. Using radio- and satellite telemetries and geolocation systems, researchers have discovered specific routes and destinations of a number of species. The chart opposite shows some examples.

In addition to its value for the conservation of migratory birds, this information is important in understanding and evaluating the spread of infectious diseases, including avian flu, within and between countries.

Species	From	To
Whooper Swan	E, N, W and C Mongolia	NE China and the Republic of Korea
Bar-headed Goose	W and C Mongolia	India
Greylag Goose	India	W Mongolia
Swan Goose	C and E Mongolia	NE China and the Republic of Korea
White-naped Crane	E Mongolia	NE China and Izumi, Japan
Demoiselle Crane	Lake Khar-Us (W Mongolia)	Kashmir-India (trans-Himalayas)
Great Bustard	N Mongolia	N and C China
Black Stork	W and C Mongolia	Myanmar and India
Curlew Sandpiper	Victoria, Australia	C Mongolia
Red-necked Stint	Victoria, Australia	Lake Buir, Ulz and Herlen Rivers in E Mongolia
Asian Dowitcher	Australia	W Mongolia
Mongolian Gull	C Mongolia	NE China and the Republic of Korea
Northern Lapwing	C Mongolia	Japan
Cinereous Vulture	C, S, W and E Mongolia	Nepal and the Republic of Korea
Short-toed Snake Eagle	S Mongolia (Gobi)	Myanmar
Pallas's Fish Eagle	C and E Mongolia	India
Steppe Eagle	C Mongolia	N China and Myanmar
Amur Falcon	C Mongolia	India and South Africa
Saker Falcon	N and C Mongolia	Yangtze River valley and Lake Höh Depression, NE China

Some examples of the origins and destinations of diurnal migrants in Mongolia

MONGOLIA AS A CENTRE OF SPECIES ORIGIN

In recent years, changes in behaviour and migration routes initiated by global climate change have resulted in many species being newly recorded in Mongolia. This is also a result of the intensive bird surveys that have been conducted by researchers in the country in recent years. According to the concept that the centre of origin of a bird species is the area in which the core breeding population is found, Mongolia emerges as one of the most important breeding grounds for several species. The country may be the centre of origin for such species as Upland Buzzard, Saker Falcon, Demoiselle Crane, White-naped Crane, Mongolian Lark, Père David's Snowfinch, Kozlov's Accentor and others.

BIRD CONSERVATION IN MONGOLIA

ENDANGERED SPECIES

Birds are under threat all over the world. According to BirdLife International, which is charged with evaluating the status of the world's entire avifauna, 1,373 species (13% or one in eight of all species globally) are threatened with extinction due to small and declining populations or ranges. Of these, 213 are Critically Endangered, meaning that they face an extremely high risk of extinction in the immediate future (Vié *et al.* 2008). While Mongolia still has a great deal of relatively undisturbed habitat and a rich avifauna, it is not immune to this alarming trend. According to BirdLife International, 7% of Mongolian bird species are Globally Threatened, a category divided under three sub-headings under which we have listed the Mongolian species at risk:

Critically Endangered: *Considered to be facing an extremely high risk of extinction in the wild.*

Baer's Pochard

Siberian Crane

Endangered: *Considered to be facing a very high risk of extinction in the wild.*

White-headed Duck

Steppe Eagle

Saker Falcon

Red-crowned Crane

Far Eastern Curlew

Yellow-breasted Bunting

Vulnerable: *Considered to be facing a high risk of extinction in the wild.*

Swan Goose

Lesser White-fronted Goose

Common Pochard

Eastern Imperial Eagle

Greater Spotted Eagle

Pallas's Fish Eagle

White-naped Crane

Hooded Crane

Horned Grebe

Dalmatian Pelican

Swinhoe's Rail

Great Bustard

Macqueen's Bustard

Relict Gull

European Turtle Dove

White-throated Bush Chat

A further category includes species that are close to qualifying or are likely to qualify for a threatened category in the near future:

Near Threatened:

Japanese Quail

Falcated Duck

Ferruginous Duck

Pallid Harrier

Cinereous Vulture

Himalayan Griffon

Bearded Vulture

Northern Lapwing

Asian Dowitcher

Bar-tailed Godwit

Black-tailed Godwit

Eurasian Curlew

Grey-tailed Tattler

Curlew Sandpiper

Red-necked Stint

Reed Parrotbill

Marsh Grassbird

Japanese Reed Bunting

In 2009, Mongolian ornithologists assessed all bird species occurring in Mongolia using the International Union for the Conservation of Nature (IUCN) Red List Criteria and categories. Of the 476 assessed native bird species of Mongolia, 10% were categorised as regionally threatened, including **Critically Endangered** (0.6%), Dalmatian Pelican and Siberian Crane; **Endangered** (1.7%), White-headed Duck, Relict Gull, Greater Spotted Eagle, Pallas's Fish Eagle, Short-toed Snake Eagle and Reed Parrotbill; **Vulnerable** (3.3%), Lesser White-fronted Goose, Baikal Teal, Ferruginous Duck, Bearded Vulture, Eastern Imperial Eagle, Saker Falcon, White-naped Crane, Hooded Crane, Asian Dowitcher, Great Bustard, Macqueen's Bustard and Henderson's Ground Jay; and **Near Threatened** (4.4%), Eurasian Bittern, Little Bittern, Purple Heron, Greater White-fronted Goose, Swan Goose, Mute Swan, Falcated Duck, White-tailed Eagle, Altai Snowcock, Common Pheasant, Common Crane, Tree Pipit, White-throated Bush Chat, Saxaul Sparrow, Yellow-breasted Bunting and Japanese Reed Bunting. Almost 90% of Mongolian birds are categorised as Least Concern (LC). Just 30 species were categorised as **Data Deficient** (DD) (Gombobaatar *et al.* 2011).

Bird conservation measures that apply in Mongolia include laws (Mongolian Law on Nature Conservation, enacted in 1995; Mongolian Law on Fauna, enacted in 2012), rules and acts governing general nature and biodiversity conservation (22 bird species are listed in the Mongolian Governmental Act No 264, issued on 5 December 2001); the Regional Red List and Conservation Action Plan for Mongolian Birds (2011) (36 species regionally threatened – Critically Endangered, Endangered, Vulnerable and Near Threatened); the Mongolian Red Book in 1997 and 2013 (36 species of birds listed in the Threatened and Near Threatened categories); the Protected Area laws and management regulations (Mongolian Law on Protected Areas, enacted in 1995); the Saker Falcon Conservation Action Plan for Mongolia (2000); and Mongolia's partnership of international conventions such as Biodiversity, Ramsar, CITES and the Convention of the Conservation of Migratory Species.

In addition to analysing and categorising global threats, BirdLife International's Endemic Bird Area Programme has mapped the global distribution of every bird species with a restricted range of less than 50,000 km^2. The areas where these ranges overlap define avian centres of endemism that are termed Endemic Bird Areas (Stattersfield *et al.* 1998). Endemic Bird Areas are excellent indicators of general biodiversity and represent priority areas for global biodiversity conservation. This programme also identifies Secondary Areas, which support one or more restricted-range bird species, but do not qualify as Endemic Bird Areas because only one of those species is entirely confined to them. There is one Secondary Area in Mongolia, defined by the range of Kozlov's Accentor, a poorly known species which breeds on rocky slopes with juniper scrub and *Caragana* bushes in the high mountains of the Mongol-Altai, Gobi-Altai and Khangai. It also occurs in arid scrub in sandy desert in the Ningxia autonomous region in northern China, where it is assumed to be a non-breeding visitor. It is not clear whether its habitat is adequately covered by existing protected areas.

Finally, Mongolia's extensive steppes are home to significant portions of the world population of several species for which, though not globally threatened, the country nevertheless has an international responsibility. These include Upland Buzzard, Demoiselle Crane, Pallas's Sandgrouse, Mongolian Lark, Père David's Snowfinch and Blyth's Pipit.

THREATS TO MONGOLIAN BIRDLIFE

Mongolia can claim some advantages in the global struggle to protect birds and their habitats. It is a very large country with a very small population, leaving lots of space for wildlife habitats. Its long-standing nomadic herding tradition has inhibited residential development, and a meat and dairy diet based on domestic animals greatly reduces the need for hunting wild game. Except for its few large cities, there has so far been relatively little industrial development. And its spectacular natural features have so far been little marred by over-use or recreational development.

However, Mongolia is changing rapidly. Almost half the population now lives in cities and only 30% of families now follow the nomadic lifestyle. The discovery of vast mineral resources, including some of the world's largest gold, copper and coal reserves in the Gobi Desert and oil in the eastern steppe, while potentially the source of great benefit to the population, also raises the spectres of pollution, poorly planned development and a newly affluent population which understandably wants to enjoy recreational opportunities in its beautiful country without perceiving the need for conservation when there is so much open space.

What follows is a list of current threats to Mongolian birdlife, along with some of the rarer species affected. It is important to note, however, that these phenomena – almost all of them involving human activity – also affect many species, which, though still relatively common, have been shown through research to be in gradual, but steady, decline (Gombobaatar *et al.* 2011).

- **Climate change** alters montane vegetation zones so that alpine species requiring barren, tundra-like habitats may be encroached upon by lusher vegetation as it moves higher. Conversely, lakes in arid steppe and desert are likely to become smaller as the climate warms – or disappear altogether.

- **Mining**. In addition to directly destroying habitats, mineral extraction typically uses large quantities of water, further contributing to droughts. And oil-drilling brings with it the expectation of spills and air pollution.

- **Dam construction** in the Great Lakes Depression in western Mongolia threatens the breeding population of the globally Vulnerable White-headed Duck and many other waterbirds.

- **Recreational development**. Lake Hövsgöl, Mongolia's 'dark blue pearl', arguably the cleanest lake in the world and certainly one of the most beautiful, has long been one of the natural world's best kept secrets. But recent improvements, such as a newly paved access road, a modernised airstrip and a proliferation of new (mostly illegal) camping facilities, have developers talking of powerboats, jet skis and luxury hotels.

- **Power lines**, particularly those of 15 kilovolts, have killed hundreds of Saker Falcons, Steppe Eagles and other raptors in the steppe.

- **Poaching**. Herders near Lake Khar-Us in the north-west continue to decimate the local population of the globally threatened Dalmatian Pelican, in order to use the upper mandible of its bill as a traditional tool for scraping sweat from their horses.

- **Illicit trade in falcons**, typically involving local poachers who sell the birds for high prices to wealthy individuals for use in falconry.

- **Large-scale steppe fires**, which are becoming increasingly frequent due to climate change, are destroying nesting habitats for rare and local grassland and reedbed species such as Marsh Grassbird, Reed Parrotbill, Japanese Reed Bunting and three species of cranes.

- **Overgrazing**. While much can be said in favour of the ecological benefits of the nomadic lifestyle, there is no question that many areas are overgrazed, resulting in the elimination of upland bird species that require a lusher, more natural steppe, or wetland species whose aquatic habitat has been degraded.

- **Threats during migration and on wintering grounds**. In addition to the natural hazards of long-distance migration, many species are subject to capture, hunting or the loss of habitat along their route or when they reach their destination. A prime example is the White-throated Bush Chat, known as a breeding bird only from Mongolia. It is mainly threatened by habitat destruction in its winter quarters in northern India and Nepal, although climate change might also have a serious impact on its alpine breeding habitat.

- **Pesticides**. Chemical treatments of pests such as Siberian Moth and Brandt's Vole cause mortality and low breeding success by entering the food web and concentrating in birds tissues.

PROTECTED AREAS

Mongolia has an extensive system of legally protected conservation habitats dedicated to preserving the country's natural and cultural heritage at the state and provincial level. This includes what may be the world's oldest nature reserve, originally designated as the Bogd Khan Strictly Protected Area in 1778. Altogether, conservation land at the state level occupies 17.4% of Mongolia's total area or about 27,250,000 ha, an area slightly larger than the state of Colorado or the United Kingdom. The system is administered at the federal level by the Department of Protected Area Administration under the Ministry of Environment and Tourism. These lands are divided into four categories based on importance and human use:

Strictly Protected Areas (SPAs)

Relatively pristine, ecologically critical wilderness areas having 'particular importance for science and civilisation'. Fourteen SPAs totalling 12,411,057 ha (representing 7.93% of Mongolian territory) have been established in Mongolia. The SPAs include three types of zone: (1) Pristine Zones, where access is allowed only for research; (2) Protected Zones, in which conservation activities such as rare species management can occur in addition to research; and (3) Limited Use Zones, in which passive tourism, traditional religious practices and some plant harvesting is allowed, but hunting, logging and construction are forbidden.

National Parks

Wilderness areas with historical, cultural or environmental education values. A total of 29 national parks, comprising 11,885,235 ha (7.6% of Mongolian territory), are actively managed. The parks are also divided into three zones: (1) Core Areas for research and conservation activities; (2) Ecotourism Zones, in which all the activities listed above are permitted in addition to fishing; and (3) Limited Use Zone in which grazing and some construction are allowed with permission from the park authorities.

Nature Reserves

There are 30 nature reserves, totalling 2,774,579 ha (1.77% of Mongolian territory). These are divided into four categories: (1) Ecosystem Reserves, protecting watersheds, habitats and other places of ecological value; (2) Biological Reserves, protecting rare species; (3) Paleontological Reserves, conserving fossil beds; and (4) Geological Reserves, preserving significant geological features. Some economic activities are allowed in nature reserves if they do not conflict with their mission.

Natural and Historical Monuments

These protect unique landscapes and cultural and historical sites. Broad public access is allowed for many uses that are non-detrimental to the site. Mongolia has 14 Natural and Historical Monuments, totalling 128,313 ha (0.08% of Mongolian territory).

Important Bird Areas

In addition to the government-designated protected areas described above, BirdLife International has identified 70 Important Bird Areas (IBAs), defined as 'sites of international importance for bird conservation at the global, regional or national level, based on standard, internationally recognised criteria'. Though the designation of an IBA confers no legal protection, it has proved to be an effective strategy for raising the profile of significant bird habitats worldwide, often leading to the conservation of areas not previously protected. Currently, Mongolian IBAs comprise 8,400,000 ha (5% of Mongolia territory), with the largest number located in Dornod, Zakhan, Khovd and Hövsgöl provinces (Nyambayar & Tseveenmyadag 2009).

CONSERVATION ORGANISATIONS AND GOVERNMENT AGENCIES IN MONGOLIA

National, Academic and Governmental Organisations

National University of Mongolia
Ikh Surguuliin Street-1, Sukhbaatar District, Ulaanbaatar, Mongolia
Tel: 976- 90112244, 976-98100148
E-mail: gomboo@num.edu.mn, mongolianbirds@mail.com
www.num.edu.mn

Mongolian Academy of Sciences
Biology Institute, Mongolian Academy of Sciences, Ulaanbaatar, Mongolia
Tel: 976-99042804, 976-91915107
E-mail: gmainjargal@gmail.com
www.mas.ac.mn

Ministry of Environment and Tourism, Mongolia
Baga Toiruu-44, Governmental Building 3, Ulaanbaatar.
Tel: 976-11-312269
www.mne.mn

Mongolian Ornithological Society
Astra building, Ulaanbaatar
P.O.Box 537, Ulaanbaatar 210646A, Mongolia
Tel: 976- 90112244, 976-99180148, 976-98100148
E-mail: info@mos.mn, mongoliabirding@gmail.com
www.mos.mn

Mongolian Bird Taxonomy and Rarity Committee
Astra building and Ikh Surguuli Street-1, Ulaanbaatar, Mongolia
P.O.Box 537, Ulaanbaatar 210646A, Mongolia
Tel: 976-99180148, 976-91915107, 976-99068399
E-mail: mongoliabirds@gmail.com
www.mos.mn

Mongolian Foundation Birds of Prey
Biology Institute, Mongolian Academy of Sciences, Ulaanbaatar, Mongolia
Tel: 976-99131174
E-mail: boogii51@yahoo.com

Birds Mongolia and Mongolica Birding
Astra building, Sukhbaatar District. Ulaanbaatar, Mongolia
P.O.Box 537, Ulaanbaatar 210646A, Mongolia
Tel: 976-88180148; 976-98100148
E-mail: info@mongolica.org, infomongolica@gmail.com
www.mongolica.org

Mongol Ecology Center (MEC) and Lake Hovsgol Conservancy
Rokmon Blgd, Suite 1001, The Constitutional St. 24, Bayangol District
P.O. Box 682, Ulaanbaatar 14250, Mongolia
Tel: 976-70100826
www.mongolec.org

Wildlife Science and Conservation Center, Mongolia
Tel: 976-70157886
E-mail: info@wscc.org.mn
www.wscc.org.mn

International Organisations

Zoological Society of London and Steppe Forward Programme
London office: Regent's Park, London, NW1 4RY, UK
Mongolia office: Biology Department, National University of Mongolia and No12, Building 25, Sukhbaatar
District, Ulaanbaatar, Mongolia. Tel: 976-99180148, 976-11321501
E-mail: Gitanjali.Bhattacharya@zsl.org, gomboo@num.edu.mn
www.zsl.org; www.zsl.org/conservation/regions/asia/mongolia

BirdLife International
The David Attenborough Building, Pembroke Street, Cambridge, CB2 3QZ, UK
E-mail: birdlife@birdlife.org.uk
www.birdlife.org

Denver Zoological Foundation
2900 East 23rd Avenue, Denver, CO 80205, USA
E-mail: zooresearch@denverzoo.org
www.denverzoo.org/conservation/overview.asp

World Wide Fund for Nature (WWF)
WWF-Mongolia office
Sukhbaatar District, Ulaanbaatar, Mongolia
E-mail: info@wwf.mn
www.panda.org

Oriental Bird Club
P.O.Box 324, Bedford, MK42 0WG, UK.
E-mail: mail@orientalbirdclub.org
www.orientalbirdclub.org

The Nature Conservancy (TNC)
Asia-Pacific Region. 4245 North Fairfax Drive, Arlington, VA 22203, USA
E-mail: asiapacific@tnc.org; eoidov@tnc.org
www.nature.org

Wetlands International
P.O. Box 471, 6700 AA Wageningen, The Netherlands.
E-mail: post@wetlands.org
www.wetlands.org

Wildlife Conservation Society (WCS)
2300 Southern Boulevard, Bronx, New York 10460, USA
Wildlife Conservation Society-Mongolia Office
P.O. Box 21, Post Office 20A, Ulaanbaatar 14200, Mongolia
Tel: 976-11323719
E-mail: eshiilegdamba@wcs.org
www.wcs.org

BIRDWATCHING IN MONGOLIA

For the adventurous birder, Mongolia offers spectacular landscapes; much unspoiled habitat; friendly and welcoming people; a high degree of safety in the cities as well as the countryside; and of course a fascinating avifauna, including a great diversity and abundance of raptors, waders, waterfowl and songbirds, especially during migration.

LOGISTICS

When to Visit

Winter temperatures hover around minus 20–25°C and not infrequently hit minus 50°C, and most of the birds have departed to their wintering grounds. Therefore the prime birding season is during the warmer months and especially during migration, from late April to mid-June and late August to early October. For best overall results, aim for the middle of these periods. Be aware that a late blizzard in spring or an early one in autumn is a distinct possibility – which you should welcome since these tend to produce big falls of migrants.

How to Get Around

If you want to sample several parts of the country (highly recommended), you will want to fly from Ulaanbaatar to a provincial capital and arrange ground transportation from there. While travel by road is rapidly improving, Mongolia is a big place with mostly dusty roads, and in many areas none at all. While it is possible to rent 4WD vehicles, you should be confident in your car repair skills if you do so, and the best option is to hire a car and driver or better still hook up with one of several Mongolian birding guides now offering their services. Also, most of the international birding tour operators, as well as local ones, offer tours of Mongolia.

Food

This used to be a challenge since the traditional Mongolian diet consists almost exclusively of meat and dairy, and preparation was, shall we say, 'simple'. But since independence in the early 1990s, the cities have an excellent selection of international cuisines to choose from and even understand about the needs of vegetarians, vegans and the gluten-averse. Any standard guide book will have a listing of eateries. Eating in the countryside is more challenging, but these days more and more Mongolians understand and cater to the tastes of foreigners. You will definitely want to try some of the local specialties like buuz and khuushuur – and if you appreciate a glass after a hard day's birding by all means try the superb local vodka.

Lodging

All of the usual international options are available: high luxury, medium-priced chain hotels, even airbnb! It just depends on how much you want to spend. In the countryside, the best option is to stay in a ger (yurt) camp, which again come in a spectrum of price ranges; the better ones have lodge buildings with hot showers, dining rooms and bars, and sometimes even offer entertainment.

Tours and Guide Services

Services catering to clients with an interest in birds or natural history in general can be arranged within Mongolia (www.mongolica.org, www.mos.mn) as well as in North America and Europe. The list of birdwatching localities below is far from comprehensive but attempts to briefly describe the most significant areas that are also reasonably accessible. Space limitations forbid giving detailed directions, but exploration and discovery are among the special pleasures of birding in Mongolia.

IMPORTANT BIRDING AREAS

Important birding areas in Mongolia

Ulaanbaatar, the capital city, is the starting point for most trips. The Tuul River, which runs through the city, and its associated gallery forest and thickets can be quite productive, especially early in the morning during spring and autumn migration. Birding the river copses near Songino in the city centre is likely to produce Black Kite, Fork-tailed Swift, Azure Tit, White-crowned Penduline Tit, Red-billed Chough, Daurian Jackdaw and Yellow-breasted Bunting, as well as a good selection of eastern Siberian migrants in season.

Twenty three kilometres to the west of the city, near Chinggis Khaan airport, small gravel ponds fed by the Tuul River and accessible by a network of raised dykes attract a variety of waders and waterfowl, as well as grassland birds. In late summer/early autumn, hundreds of Ruddy Shelducks congregate here.

The Bogd Khan Strictly Protected Area (41,651 ha), just south of the city, is the southernmost extent of the Hentii mountain taiga (boreal forest), and contains a mixture of coniferous and deciduous forest as well as scrub, steppe and boulder fields at its highest points. More than 150 species of birds have been recorded here, including Amur Falcon, Eurasian Three-toed and Black Woodpeckers, Hill Pigeon, Blyth's, Richard's and Olive-backed Pipits, Thick-billed and Dusky Warblers, Brown Shrike and Pine Bunting. In winter, Common Redpoll and Long-tailed and Pallas's Rosefinches are found in some numbers here.

Gorkhi-Terelj National Park (293,168 ha). Reached by paved road 54 km north-east of Ulaanbaatar, the park borders the Khan Khentii Strictly Protected Area (1.2 million ha) and includes the southern Hentii mountain range. The park offers a variety of habitats such as coniferous and mixed forest, rocky outcrops, rich meadows, small streams bordered by broadleaved woodland and mature riparian poplar forests along the Tuul River. The bird fauna is similar to that of Bogd Khan Uul but many more species have been recorded. The most interesting species for birders are breeding Eastern Buzzard, all the woodpeckers and tits, Siberian Rubythroat, Red-flanked Bluetail, Daurian Redstart, Chinese Bush Warbler, Dark-sided and Taiga Flycatchers, Red-throated Thrush, Pine Bunting and the rare Black-billed Capercaillie.

Hustai National Park (50,600ha). Driving west on the main paved road out of Ulaanbaatar, watch for Upland Buzzard, Cinereous Vulture, Demoiselle Crane and assorted lark and wheatear species in the open steppe.

The dirt access road into the park is signposted about 95 kms west of the capital. Hustai is best known as a reintroduction site of the Takhi or Przewalskii's Horse, which became extinct in the wild in the late 1960s. However, it also contains an excellent variety of habitats including natural steppe, montane forest and the riparian habitat along the Tuul River. Some of the park's specialities include breeding Daurian Partridge, Black Stork, Golden, Steppe and Booted Eagles, Amur and Saker Falcons, Pied and Isabelline Wheatears, Mongolian and Asian Short-toed Larks, and Meadow Bunting. The gift shop stocks a photographic guide to the birds of the park as well as to its other wildlife and plants.

To the west of Hustai National Park (*c*. 400 km west of Ulaanbaatar), the freshwater steppe lake of **Ögii Nuur/Lake** encompasses a rich and fairly accessible wetland, where White-tailed and Pallas's Fish Eagles, Swan and Bar-headed Geese, Whooper Swan and many other species can usually be seen. There are also several records of White-naped, Hooded and Siberian Cranes.

Gun Galuut Nature Reserve. Just 2½ hours (130 km east) from Ulaanbaatar, this newly established 20,000 ha-reserve contains an extraordinarily beautiful river valley and mountain range surrounded by rolling steppe. Marshes near the comfortable ger camp hold nesting Whooper Swan and White-naped Crane, and a chain of small lakes host myriad grebes, waterfowl, waders (e.g. godwits, Asian Dowitcher, Spotted Redshank), gulls and terns on migration. Riparian willow copses are home to White-crowned Penduline Tit and migrant warblers. There is also a small population of Argali Sheep in the surrounding rocky hills.

Kharkhorin (Karakorum), the Mongol capital in the 13th and 14th centuries, lies *c*. 370 km to the south-west of Ulaanbaatar. A trip here can easily be combined with a visit to Ogii Nuur (see above). Close to Kharkhorin, the largest agricultural area in Mongolia attracts congregations of up to 7,500 Demoiselle Cranes in September, and Eurasian Eagle Owl and Blyth's Pipit have been recorded breeding in the buildings of the nearby Buddhist monastery of Erdene-Zuu.

Khangai Mountains in central-western Mongolia are one of the country's major ranges with the highest peak (Otgontenger) reaching 4,021 m. The range contains a wealth of undisturbed habitat from forest steppe at lower elevation and larch forest on north-facing slopes to alpine meadows. Near the town of Uliastai the little-known White-throated Bush Chat occurs in subalpine meadows near streams, in areas with gorges, rocky outcrops and scattered boulders, at an altitude of 2,430–2,600 m. It shares this habitat with Altai Accentor and other high-altitude species.

Hövsgöl National Park (838,070 ha). Lake Hövsgöl, lying at the northernmost point in Mongolia, is a high (1,645 m), long (137 km), deep (c 250 m) rift lake, a little sister to Lake Baikal, into which it drains *c*. 200 km to the north-east, and with which it shares many similarities in origin, flora and fauna. A key difference, however, is that Hövsgöl has suffered very little development to date and is therefore one of the cleanest lakes in the world. The lake is surrounded by a vast national park of 1.2 million hectares – larger than Yellowstone. The forest avifauna is comparable to that of the Hentii and Khangai Mountains. The best birdwatching areas lie in the southern half of the western shore, accessible by paved and gravel roads and the (less accessible) Alagtsar River valley along the eastern shore. In addition to the lake itself, the habitats within the park include, lagoons and wet meadows, mature larch forest and alpine barrens at the highest elevations. This combination of habitats has attracted more than 250 bird species. In early to mid-June, when the main lake is typically still frozen, waterbirds congregate in numbers in and around shoreline lagoons and leads in the melting ice. They include Black-throated Diver, Horned Grebe, Whooper Swan, Bar-headed Goose, Ruddy Shelduck, White-winged Scoter, Common Goldeneye, Northern Lapwing, Little Ringed Plover, Common Redshank, Long-toed Stint (rare breeder), Mongolian and Black-headed Gulls, and Common and White-winged Terns. White-tailed Eagles are a frequent sight near these food sources, and three species of wagtails (including Citrine), patrol the water's edge. Typical birds of Hövsgöl's forests and forest edges include Black-billed Capercaillie, Eurasian Hoopoe, up to six species of woodpeckers, Oriental Cuckoo,

Tree and Olive-backed Pipits, several species of leaf warblers and tits, Eurasian Nuthatch, Dark-sided and Taiga Flycatchers, Brambling, Red Crossbill (sporadic) and Pine Bunting. Swinhoe's Snipe breeds locally in boggy forest openings, and the region is one of the few places outside Siberia where Black-throated and Red-throated Thrushes breed alongside each other.

Erkhel Lake. If you are driving between Lake Hövsgöl and the provincial capital of Mörön, you will come upon a shallow lake in a grassy valley about 40 km from the town. This is often filled with waterfowl and other aquatic species, and the surrounding steppe is filled with larks and pipits.

Uvs Lake Basin Strictly Protected Area. Uvs Nuur is Mongolia's largest lake by surface area (771,000 ha) and is surrounded by spectacular sand dunes and snow-capped mountains. It is a shallow, saline lake with an extraordinary diversity and abundance of waterfowl, wading birds, shorebirds, and wetland and steppe songbirds. Among its birds of special interest are Swan Goose, Red-crested Pochard, a colony of Eurasian Spoonbills, Black Stork and Pallas's Gull. Relict Gull has also been recorded here. The lake shore is readily accessible from the nearby provincial capital of Ulaangom. A visit to Uvs is easily combined over several days with an exploration of the Great Lakes Depression (see below).

The Great Lakes Depression in western Mongolia contains some very large lakes: Khyargas (1,407 km^2; saline), Airag (143 km^2; freshwater), Khar-Us (1,852 km^2; freshwater), Khar (575 km^2; freshwater) and Dörgön (305 km^2; saline). All of these lakes hold species characteristic of the region, including Bar-headed Goose and other waterfowl, the eastern subspecies of Black-tailed Godwit (*L. l. melanuroides*), and many other waders, Pallas's Gull, and White-winged, Whiskered, Little, Caspian and Gull-billed Terns.

Khar-Us Nuur National Park (850,272 ha) has some of the largest reedbeds in Central and Eastern Asia. They support breeding Eurasian Bittern, Bearded Reedling, Paddyfield, Great Reed and Blyth's Reed Warblers, Common Reed Bunting and the beautiful white-headed subspecies of Yellow Wagtail (*M. f. leucocephala*). Another speciality here is the world's easternmost population of the globally endangered White-headed Duck.

Airag Nuur may be the most bird-rich wetland in Mongolia. Especially in late summer and early autumn, tens of thousands of birds use the area as a stop-over site. A wide variety of waders can be observed feeding on exposed mud when water levels are low, especially at the mouth of the Zavkhan River, e.g. Greater Sand Plover, Broad-billed and Sharp-tailed Sandpipers, and Asian Dowitcher. Waterfowl are plentiful, with Red-crested Pochard, Eurasian Wigeon and Eurasian Teal occurring in thousands and providing an abundant food source for White-tailed Eagles and Pallas's Fish Eagles, which are present in dozens in late summer and early autumn. Besides a large colony of Eurasian Spoonbills, the lake regularly holds a few of the globally vulnerable Dalmatian Pelican, a species that may have recently ceased to breed in Mongolia due to persecution.

The Gobi. Travel to desert regions requires expedition-like logistics and is best arranged through a reliable travel operator. Many ger camps now provide comfortable (even luxurious!) lodging for exploring this vast and beautiful wilderness. Key Gobi species include Macqueen's Bustard, Pallas's Sandgrouse, Oriental Plover, Henderson's Ground Jay, Kozlov's Accentor, Asian Desert Warbler, Desert Wheatear and Saxaul Sparrow, plus many montane species (see below).

Dalanzadgad, where most visitors arrive in the Gobi by air from Ulaanbaatar, has a leafy central avenue and some agricultural fields (Bayandalai) with a windbreak of poplar trees that can be full of migrants, sometimes including rarities, attracted to these islands of green in an otherwise arid landscape. On a good day it is possible to see scores of warblers, chats, thrushes and flycatchers feeding in these groves and adjacent fields, often chased about by migrant accipiters or falcons. There is also a wetland area to the east of the city with hides catering for birdwatchers.

Gobi Gurvansaikhan National Park (2,694,737 ha), situated north-west of Dalanzadgad, encompasses a series of ranges and valleys that make up the eastern end of the Gobi-Altai mountains, which rise 2,200–2,600 m above the surrounding plains. The park, which extends for more than 380 km from east to west, lies at the northern edge of the driest part of the Gobi but the park's landscape is extremely varied, with rocky and sandy desert plains, arid steppe, precipitous cliffs and ravines, patches of juniper, saltpans and oases. Over 240 bird species occur in the area. An essential stop for birders exploring the Gobi is **Yolyn Am** (Valley of the Bearded Vulture). True to its name, this spectacular gorge is also home to numbers of Cinereous and Himalayan Vultures, as well as Bearded Vultures, which fill the skies as they leave their roosts at mid-morning to explore the plains. However, the valley is also rich in many other interesting birds such as Altai Snowcock (at the highest elevations), Wallcreeper, Black and Güldenstädt's Redstarts (the latter at high altitude), Alpine, Brown and Kozlov's Accentors, Mongolian Finch, White-winged Snowfinch, Himalayan Beautiful Rosefinch and Godlewski's Bunting. Wild Argali Sheep and Asiatic Ibex are also seen in Yol and a few lucky birdwatchers have even glimpsed the elusive Snow Leopard here.

The Valley of the Lakes. A number of brackish Gobi lakes, most notably Bööntsagaan, Orog, Taatsyn Tsagaan, Holboolj and Ulaan Nuur lie in a broad valley north of the Gobi-Altai range and south of Khangai. During migration periods, these lakes can hold an abundance of birdlife similar to that of the Great Lakes Basin (see above). These are the most reliable sites to find the unpredictable and globally threatened Relict Gull.

The Eastern Steppe of Mongolia's Dornod province is the largest relatively undisturbed grassland ecosystem in the world. It is ten times the size of the Serengeti and contains the largest herds of hoofed animals outside of Africa. Herds of Mongolian Gazelles, often numbering in the tons of thousands, still breed in and migrate across this vast ocean of grass. The grasslands also provide habitat for rare resident bird species such as the globally threatened Great Bustard, as well as for thousands of northward-bound Pacific Golden Plovers and, especially in autumn, large numbers of Little Curlews.

Choibalsan, Dornod's provincial capital, 600 km east of Ulaanbaatar, can be reached by fairly good roads in a day and a half. Like Dalanzadgad (see above), plentiful poplars in the town create an oasis-like effect and during spring migration this green island can be bustling with migrants, including hundreds of Taiga Flycatchers, Siberian Blue Robins, the occasional White-throated Rock Thrush and their attendant predators.

Dornod also contains numerous lakes and river drainages, with associated reedbeds and other wetlands, such as the **Herlen River** and the small lake system of **Tashgain Tavan Nuur**, which due to their geographical location results in frequent records of East Asian species at the western edge of their range, such as Oriental Pratincole, Chinese Pond Heron, Reed Parrotbill, Marsh Grassbird and Japanese Reed Bunting.

The Mongol Daguur Strictly Protected Area (103,016 ha) was established to preserve a representative portion of Mongolia's Daurian steppe and its characteristic flora, fauna, landscape and endangered species. Six species of cranes have been recorded: Common (passage and breeding), Hooded (common on passage, especially in autumn), White-naped (common breeder), Red-crowned (very rare summer visitor), Siberian (rare non-breeding summer visitor and passage migrant) and Demoiselle (common breeder). The protected area is especially known for its high density of nesting White-naped Cranes; the Ulz River, which flows through the southern part of the reserve, and the nearby Onon and Khurkh rivers support one of the largest breeding populations of this species in the world.

The Nömrög Strictly Protected Area (311,205 ha) covers the remote and uninhabited far eastern tip of Mongolia and includes the westernmost end of the Khyangan mountain range, which extends into Mongolia from Manchuria, China. Mandarin Duck, Yellow-rumped Flycatcher and Black-naped Oriole, inhabiting riparian forest, reach the westernmost limit of their breeding ranges here. As it is situated near the border, a special permit to visit this area is required.

Send us your records

Given Mongolia's size and the relatively small size of its birding community, there is a great deal still to be learned about its bird distributions. You can help by sharing your records with us. The ideal record includes exact date and place of your observation, detailed field notes on plumage (age and sex), habitat and conditions, and a photo or field sketch of the species in question. We would also be grateful for locality lists from the places you visit. Send your communications to the **Mongolian Ornithological Society** (Tel: 976- 90112244, 976-98100148. E-mail: info@mos.mn, mongoliabirding@gmail.com. Website: www.mos. mn), **Mongolian Bird Taxonomy and Rarity Committee** (Tel: 976-99180148, 976-91915107, 976-99068399. E-mail: mongolianbirds@mail.com, boogii51@yahoo.com. Website: www.mos.mn) and **National University of Mongolia** (Tel: 976-90112244, 976-98100148. E-mail: gomboo@num.edu.mn, mongoliabirds@gmail.com. Website: www.num.edu.mn).

BIRDING CODE OF CONDUCT

Everyone who enjoys birds and birding must always respect wildlife, its environment, and the rights of others. In any conflict of interest between birds and birders, the welfare of the birds and their environment comes first. For more information about the American Birding Association's Code of Birding Ethics please go to www.aba.org.

Relict Gull, one of the world's rarest gulls (*G. S.*)

HOW TO USE THIS BOOK

FINDING AND IDENTIFYING BIRDS

Where and when to look

Birds are everywhere, but different species prefer to live in different kinds of physical surroundings or habitats such as forests, grasslands or wetlands. Some species are typically restricted to a single habitat, while others are more wide-ranging. To see the greatest variety of birds, the best strategy is to visit as many habitat types as possible. Over millennia birds have also evolved unique distribution patterns or geographical ranges, so that different species will be present or absent from different parts of the country. Different species also occur at different times of year, some remaining year-round, some visiting during the nesting season and then migrating elsewhere for the winter, while others pass through during spring and autumn without nesting or appear only in winter. The species accounts in this guide note the preferred habitats and seasons in which each species is likely to be present in Mongolia, and the range maps show geographical distribution and seasonality. See also Birdwatching in Mongolia, p. 30 for the country's birding hotspots.

Equipment

Some birds are so distinctive that they can be easily identified almost as far away as they are visible. But one of the enticing challenges of birdwatching is that many closely resemble one or more other species and therefore need to been seen well enough to discern distinctive details. Since birds generally keep their distance from people, a good pair of binoculars is practically essential. Binoculars are categorised by a pair of numbers indicating the degree of magnification and the breadth of field (how broad an area is visible). Birdwatchers typically use binoculars with a magnification value between 8 and 10 and a field of 30–40. Binoculars come in a wide range of quality and price, but good optics are very important for the quality of the birding experience, so a useful rule of thumb is to buy the most expensive pair you can afford. If possible, it is advisable to try out several different brands to evaluate criteria such as weight and ease of focusing.

For identifying birds at a distance or when subtle field marks need to be distinguished, e.g. with confusing species of waders (shorebirds), a spotting telescope can be extremely useful – and also allows increased appreciation of plumage colours and patterns in even common birds.

'Serious' birders have traditionally carried a field notebook in which to record plumage characteristics in writing and sketches for reference in cases of hard-to identify species or rarities for which documentation is desirable. Increasingly, notebooks are being replaced by lightweight cameras capable of capturing excellent images at high magnification without cumbersome lenses.

Bird topography

An effort has been made in this guide largely to avoid technical terms. However, a number of avian anatomical features and feather groups are useful in describing a bird's characteristics. These are illustrated and labelled on pp. 39–40.

Variations in appearance

A single bird species can appear in strikingly different variations depending on season, state of moult and sex. This guide shows standard variations in different ages and sexes as well as flight patterns where useful. However, birders also need to be aware of other factors that can make a significant difference in a bird's appearance. These include: (1) the degree of feather wear because birds generally renew their entire plumage every year, altering their appearance in different stages of this moulting process; (2) posture, since the same bird can look taller or shorter, fatter or thinner, depending on its behaviour; and (3) lighting and other prevailing atmospheric conditions.

Birding by ear

Experienced birders depend to a great extent on the sounds birds make to both locate and identify them. All birds make sounds unique to their species for a variety of purposes such as attracting a mate, proclaiming a breeding territory, expressing distress (e.g., the presence of a potential predator) or promoting cohesion within a flock. Knowing bird songs and calls is especially useful when looking for reclusive species in dense habitats such as forests or marshlands, but it can also clinch an identification in species that look nearly identical but sound very different. Voice cues are most useful of course during the breeding season when males of many species advertise their presence constantly to females and rival males of their species.

This guide includes voice descriptions (songs and calls) for all species, and these may be useful in making gross distinctions, for example, between a buzz, a chirp and a warble. But sounds are difficult to convey precisely in words. Fortunately, songs of most of the world's bird species have now been recorded and can now be downloaded to your smartphone (see for example www.xeno-canto.org). However, the most effective – and enjoyable – way to learn bird songs is to chase down and identify the singer.

Bird behaviour

In addition to identifying birds and keeping lists of the species you've seen, paying attention to distinctive behaviours can contribute greatly to the pleasures of birding. The raptor or tern that seems to be attacking you, is almost certainly defending its eggs or young in a nest nearby. Why do larks and other open country birds sing while hovering high in the air? Because they lack the tree-top perches that forest birds use to proclaim their territories and attract a mate. Why do eagles, vultures, gulls and other birds often fly in ascending circles overhead? They are being lifted by columns of warm air called thermals, which allows them to gain altitude and drift off in the desired direction without expending much energy. The most interesting of these behaviours are noted in the species accounts.

KEY TO THE SPECIES ACCOUNTS

Classification and nomenclature

All species names are given in English, along with the scientific name. With some exceptions, taxonomy as well as scientific and vernacular nomenclature follows IOC World Bird List, Version 8.2 (2018). Some alternative names are given at the end of the species accounts after the subheading 'Alt'.

Measurements

Overall length (tip of bill to tip of tail) is given in centimetres.

Identification (ID)

Describes species based on their most distinctive characteristics and comparison to other species with which they are most likely to be confused, rather than detailed bill-to-tail descriptions.

Voice

Describes a species' characteristic breeding/territorial song, followed by distinctive flight or contact calls.

Habitat

Describes the species' preferred ecological community based on vegetation (e.g., coniferous forest, arid steppe) or wetland type (e.g., bog, saline lake) both during and outside of the breeding season.

Behaviour

Describes distinctive behavioural traits as relating both to identification and general interest.

Status

Describes species' (1) abundance, (2) breeding status, (3) distribution and (4) seasonality. Generally, vagrant species are included in the species accounts, but very recent vagrants are listed separately at the end of the species accounts.

Conservation

In cases of conservation concern, describes global status based on the criteria and analysis of BirdLife International, as well as regional or local conservation status where applicable. If no criterion is included in a species account, that species is considered to be of Least Concern by BirdLife International.

Taxonomy

In cases of recent taxonomic change, the former taxonomic treatment is detailed.

Abbreviations

The text also includes the following abbreviations:

ad – Adult
Alt – Alternative name of a species
cf – compare
IBA – Important Bird Area
imm – Immature
IOC – International Ornithological Congress (checklist – see p. 37)
juv – Juvenile
N, S, E, W and C Mongolia -– Northern, southern, eastern, western and central Mongolia
NGO – Non-governmental Organisation
vs – versus, used here when making a comparison between two or more species

KEY TO THE RANGE MAPS

The range maps are generalised indications of species distributions in Mongolia based on ecological zones and habitats (see map on p. 13), topographical regions (map, p. 13) or in the case of species with a very limited distribution, a spot location. The location of Ulaanbaatar is shown as a small black square on all maps.

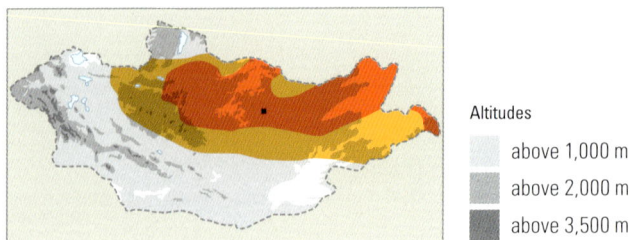

Altitudes

above 1,000 m
above 2,000 m
above 3,500 m

Maps are colour-coded according to the following criteria:

Resident year-round (incl. breeding)   Migration range (neither breeds nor winters)

Breeding/summer range   X   Vagrant

Winter resident/visitor

TOPOGRAPHY

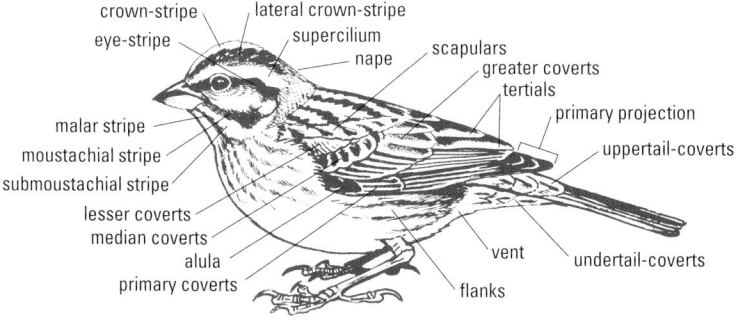

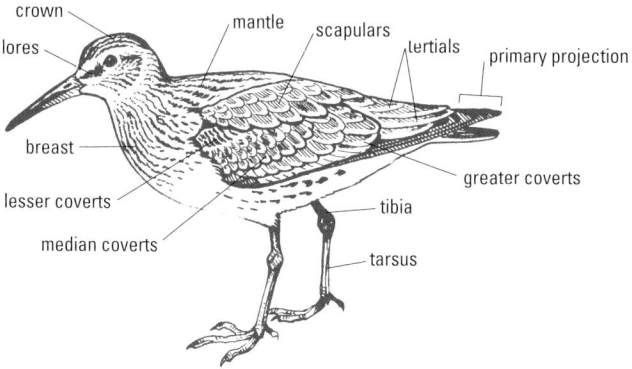

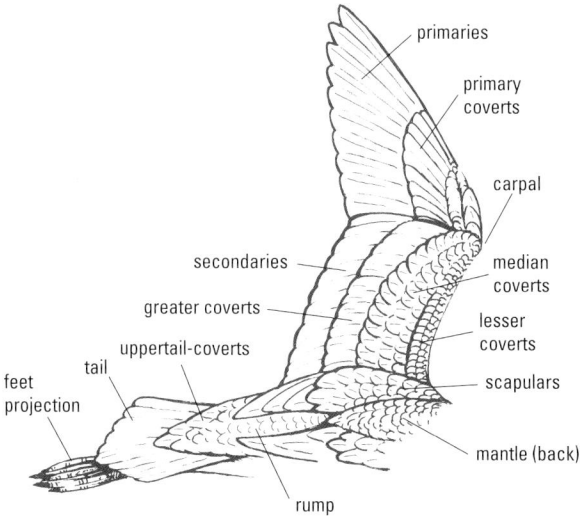

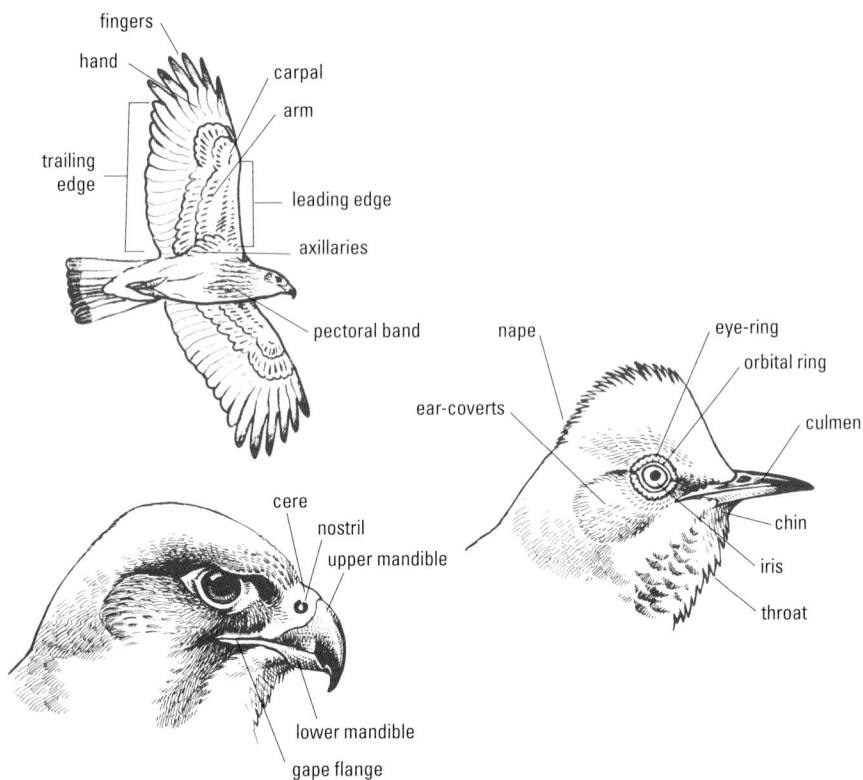

fingers

hand

carpal

arm

trailing edge

leading edge

axillaries

pectoral band

nape

eye-ring

orbital ring

ear-coverts

culmen

cere

nostril

upper mandible

chin

iris

throat

lower mandible

gape flange

GLOSSARY

Accidental: *see* Vagrant

Altitudinal migrant: species that breeds at high altitudes (in mountains) and moves to lower levels and valleys after the breeding season.

Arboreal: tree-dwelling.

Axillaries: feathers in the 'armpit' at the base of the underwing; distinctive coloration can be useful in identifying certain species in flight.

Cap: well-defined patch of colour (or bare skin) on the top of the head.

Carpal (or **carpal joint**): bend of the wing.

Carpal patch: area of contrasting colour at the carpal joint on the underside of the wing (e.g. in buzzard species).

Cere: fleshy (sometimes brightly coloured) structure at the base of the bill containing the nostrils; distinctive in some groups, e.g. diurnal raptors.

Collar: well-defined band of colour that encircles or partly encircles the neck.

Colonial: referring to bird species that typically nest close together in same- or mixed-species assemblages.

Conspecific: belonging to the same species, typically used in comparing distinctive subspecies.

Crest: tuft of feathers on crown of head, sometimes erectile.

Eclipse plumage: female-like plumage acquired by males of some species (e.g. ducks) during or after breeding.

Endemic: refers to species whose distribution is restricted to a specific country, region or topographic feature.

Eye-ring: a narrow area of contrasting feathering or bare skin surrounding the eye.

Flight feathers: primaries, secondaries and tail feathers as a whole.

Fringes: contrasting (usually pale) feather margins that frequently result in a scaly appearance to the body feathers or coverts.

Frugivorous: fruit-eating.

Gape: fleshy margin of a bird's mouth at the base of the bill, sometimes extending to below the eye.

Granivorous: feeding on grain or seeds.

Gregarious: living in flocks or colonies.

Immature (imm.): referring to plumages of young birds; non-adult plumages.

Juvenile (juv.): referring to the first set of true feathers acquired by young birds; often distinct from plumages of older birds of the same species.

Leading edge: front edge (usually of the forewing in flight).

Local: occurring within a restricted area or areas.

Mandibles: two halves of a bird's bill – upper mandible and lower mandible.

Mask: dark area of plumage surrounding the eye and often covering the ear-coverts.

Mirrors: describes contrasting white spots in the black wingtips of some gull species.

Morph: a distinct plumage variation, e.g. the dark and light 'morphs' of Rough-legged Buzzard.

Nomadic: occurring unpredictably typically depending on food or habitat availability as in irruptive finch species.

Nominate: referring to the first-named subspecies of a species, which repeats the specific name (e.g. *Anser tabalis fabalis* cf. *Anser fabalis middendorfi*).

Plantation: group of trees (often non-native species) planted in close proximity to each other for timber, as a wind-break or as a crop.

Primary projection: extension of the primaries beyond the longest tertial on a closed wing; this can be of critical use (e.g. in the identification of larks or *Acrocephalus* warblers).

Race: *see* Subspecies.

Speculum: iridescent panel in the secondaries of many dabbling ducks, often bordered by pale edges above and below.

Spectacles: refers to a contrasting incomplete ('broken') eye-ring, distinctive in some birds, e.g. Pallas's Gull.

Subspecies (also called race): geographical population whose members show consistent differences (e.g. in plumage) from other populations of the same species.

Terminal band: contrasting dark or pale band at the end of the tail.

Terrestrial: occurring mainly on the ground.

Trailing edge: rear margin of the extended wing, sometimes useful in flight identification when contrasting in colour, e.g. in snipes and skylarks.

Undertail-coverts: area of feathers that extends from the vent to below the base of the underside of the tail; the comparative length of the undertail-coverts can be useful in identification.

Underwing-coverts: area of feathers that covers the base of the flight feathers on the under surface of the wing (also known as 'wing linings'); the colour of this area of feathers can be useful in identification, e.g. the contrasting white underwing-coverts of Amur Falcon.

Vagrant: very rare and irregular in occurrence; essentially synonymous with 'accidental'.

Vent: area around the cloaca (anal opening), just behind the legs (not to be confused with the undertail-coverts).

Window: contrasting pale area at the base of the primaries, distinctive in some gulls and raptors.

Wing-bar: generally a narrow and well-defined dark or pale bar across the wing, and often refers to a band formed by the pale tips to the greater or median coverts (or both as in double wing-bar).

Wing stripe: contrasting, usually pale, narrow band at the margin of the flight feathers and the wing-coverts, often useful in the identification of waders.

Wing-panel: contrasting band of colour across some part of the wing above; broader and typically more diffuse than a wing-bar.

Willow Grouse *Lagopus lagopus*   36–43 cm

ID This species and Rock Ptarmigan are stocky grouse of open habitats, distinguished from all other Mongolian grouse by their white wings and underparts. In winter, both species turn white except for contrasting black tail feathers (best seen in flight). Willow Grouse differs from similar Rock Ptarmigan by larger overall size, heavier bill and (in spring and summer) reddish-brown rather than mottled grey plumage above. Habitat also differs (see Rock Ptarmigan). **Voice** Loud, harsh *go-back go-back go-back* and assorted guttural barking calls from male in display. **Habitat** Heathy shrublands in subalpine zone of higher mountains; typically occurs at lower elevations than Rock Ptarmigan. **Behaviour** Gregarious, especially in winter, when flocks may move to lower elevations. Feeds and nests on ground. **Status** Uncommon to fairly common but local resident breeder in all of Mongolia's major mountain regions. [Alt: Willow Ptarmigan]

Rock Ptarmigan *Lagopus muta*   33–38 cm

ID Slightly shorter-bodied and longer-tailed than Willow Ptarmigan, with mottled grey rather than reddish-brown upperparts in spring and summer and shorter, thinner bill; in white winter plumage, area between eye and bill (lores) black. For distinctions from other grouse, see Willow Grouse. **Voice** Throaty *kuh kuh gwuah* croaks from both sexes. Male gives rattling cackle in display. **Habitat** Alpine barrens of high mountains, normally occurring at higher altitudes than Willow Ptarmigan. **Behaviour** Forages and nests on ground. Gregarious, especially in winter. Sometimes very approachable. Male performs display flight (accompanied by rattling call) consisting of rapid climb on fast-beating wings then slow glide back to ground. **Status** Uncommon and local resident breeder at high elevations in Altai, Khangai and Hövsgöl regions, but very rare further east. [Alt: Ptarmigan]

Black Grouse *Lyrurus tctrix*   M *c.*60 cm, F *c.*45 cm

ID Medium-sized grouse with small head and bill. Male confusable only with two species of capercaillies, which are much larger and heavier-billed with rounded (vs. lyre-shaped) tail. In flight, male Black Grouse shows prominent white wing-stripe. Female distinguished from similar capercaillies by smaller size and bill, little or no rufous on throat and tail and narrow white wing-stripe in flight. Female Willow Grouse has white wings, belly and leg feathering. **Voice** Displaying male gives a bizarre bubbling croon *kru kru kru*, and whooshing *chooeeesh* sounds. Female cackles, *kakakakaka*. **Habitat** Forest openings, river valleys, bogs and heathlands in northern taiga and mountains. **Behaviour** In early spring, males 'dance' in open communal display grounds (leks). Forages in trees and on ground. Sociable outside breeding season. **Status** Fairly common resident breeder in northern and mountain forested regions, but often elusive outside breeding season.

Hazel Grouse *Tetrastes bonasia*   35–34 cm

ID Small, beautifully patterned forest grouse. Crested head, densely spotted underside and black and white terminal tail-band distinguish it from all other Mongolian grouse. Female is less boldly patterned, with a smaller crest and lacks male's distinctive black throat and white 'moustaches'. **Voice** Loud, shrill, sibilant songbird-like *tsee tsee tseeee*. **Habitat** Prefers dense, mature forest in mountains and river valleys. **Behaviour** Very unobtrusive but often tame when encountered. Nests on ground, but often feeds and roosts in trees. **Status** Fairly common resident breeder in mountain ranges of the north and far east, though absent from most of the Altai. **Taxonomy** Formerly placed in the genus *Bonasa*.

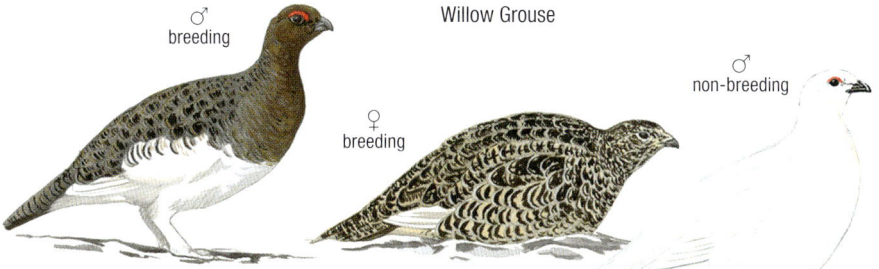

♂ breeding

Willow Grouse

♀ breeding

♂ non-breeding

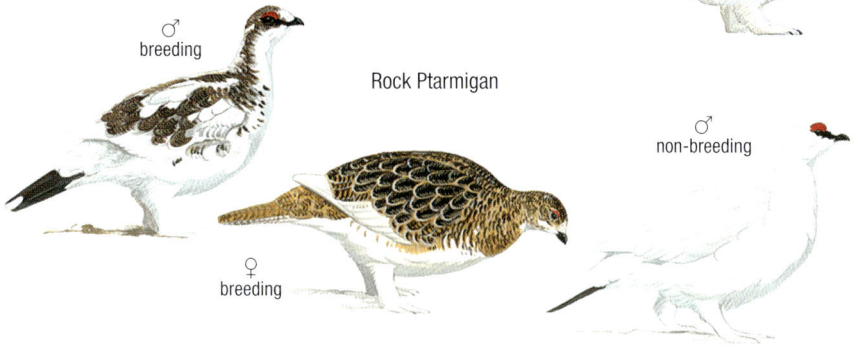

♂ breeding

Rock Ptarmigan

♂ non-breeding

♀ breeding

♂

Black Grouse

♀

♀

♂

Hazel Grouse

♂

♀

♂

Western Capercaillie *Tetrao urogallus*   M 74–90 cm, F 54–63 cm

ID Very large, turkey-like grouse, with large head and full, rounded tail, which may be raised and fanned. Size alone distinguishes this species from all other grouse except similar Black-billed Capercaillie, from which male may be told by its yellow bill, white belly and lack of white spotting on wings and tail. Smaller, brown female distinguished from female Black-billed by more rufous coloration, clear (unbarred) reddish breast and absence of white barring on wings. **Voice** In display, male makes assorted bizarre sounds: knocking, drumming, popping and wheezing. Generally quiet otherwise. **Habitat** Deciduous and mixed forest. **Behaviour** Feeds mainly on ground but often perches in trees. Normally very shy, though male at start of breeding season can become bold, even aggressive. In early spring, males gather in leks at dawn and display for females. **Status** Rare and local resident in Hövsgöl region; breeding not confirmed.

Black-billed Capercaillie *Tetrao parvirostris*   68–97 cm

ID This species and much rarer Western Capercaillie are distinguished from all other Mongolian grouse by their great size and robust (turkey-like) proportions. Male lacks yellow bill and white belly of Western Capercaillie and has prominent white spotting on wings and tail-coverts. Female is greyer (less rufous) than Western, with heavily barred (not clear) breast and conspicuous white spotting on wings. **Voice** In display, male gives an accelerating series of clacking sounds. **Habitat** Larch and pine forests of taiga region. **Behaviour** Nests and feeds on ground. Forms large flocks in autumn, which rarely move down to river valleys and wheat fields in winter. Males display for females at communal leks. **Status** Uncommon resident breeder in northern mountain regions: Khangai, Hövsgöl and Hentii. **Conservation** Widespread in northern Asia, but subject to concerns about poaching, habitat destruction and other factors. Status poorly understood and not presently categorised as a species of concern. [Alt: Spotted Capercaillie]

Altai Snowcock *Tetraogallus altaicus*   57 cm

ID Large, stocky, pale smoky greyish-brown grouse with contrasting white underparts. Cannot be mistaken for any other member of its family in Mongolia. Sexes similar. **Voice** A harsh, whistling *gyuk-gyuk-gyuk* notes gradually shortening and ending with a rolling *rrruuuuu*. **Habitat** High-altitude rocky slopes and alpine meadows. **Behaviour** Forages on ground for plant material and insects. Family groups remain together after nesting and combine into winter flocks. **Status** Uncommon and local resident breeder at high elevations in Altai (including Gobi ranges), Khangai and Hövsgöl regions; absent from Hentii. **Conservation** Listed as Near Threatened in the Mongolian Red Book (2013).

Chukar *Alectoris chukar*   32–39 cm

ID Red bill and legs, clear white throat bordered by bold black border and black flank stripes render this medium-sized grouse unmistakable in Mongolia. Sexes alike. **Voice** A series of chuks, ending in *chukARR, chukARR*; also a soft pitch *whitoo* when flushed, and males perform rapid chorus *ka ka ka ka-ka-ka-ka-kakakakaka*. **Habitat** Dry, rocky outcrops and open, rocky mountain areas with sparse vegetation. **Behaviour** Forages on ground for plant material and insects. Prefers to run when disturbed rather than flying, then 'explodes' noisily when pressed. **Status** Fairly common resident breeder on arid slopes and valleys of Mongol-Altai, Gobi-Altai and Khangai mountain ranges.

♀

♂

Western Capercaillie

♂

Black-billed Capercaillie

♂

♀

Chukar

♂

Altai Snowcock

Daurian Partridge *Perdix dauurica*   *c*.30 cm

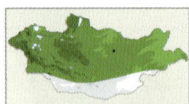

ID Orange-rufous face and throat, prominent flank stripes and black belly blotch of adult male are diagnostic. Adult female is similar but duller and less strongly marked, and juv is duller still. In winter, adult male develops 'whiskers' and 'beard'. **Voice** Assorted rasping or creaking calls; also, accelerating *rex rex rex* call. **Habitat** Dry rocky areas, river beds and steppes with grasses and low shrubs in mountains and foothills. **Behaviour** Typically occurs in close flocks of five or more birds, more so in winter. Prefers to run from danger, flushing only as a last resort. Nests and forages on ground, feeding on plant material and insects. In harsh winter conditions, visits nomad camps to pick seeds from livestock dung. **Status** Common resident breeder at lower elevations throughout highlands of Mongolia.

Common Quail *Coturnix coturnix*   16–18 cm

ID This and Japanese Quail are smallest (Common Starling size) gallinaceous birds in Mongolia with streaky nondescript plumage and pointed wings in flight. Adult male in breeding plumage has diagnostic black 'anchor' mark on throat in contrast to clear rufous throat of breeding male Japanese Quail. However, non-breeding male and adult female plumages of the two species are very similar and often not safely distinguishable in field. Compare also Yellow-legged Buttonquail and Corncrake. **Voice** Male call is a far-carrying liquid *whit, whit-tit*, often rendered as 'wet my lips' and repeated in quick succession. **Habitat** High grasses and other densely vegetated open habitats, typically in river valleys. **Behaviour** Unobtrusive and skulking, more often heard than seen. When flushed, small size, bullet shape (pointed at both ends) and pointed wings are distinctive in both *Coturnix* species. Nests and forages on ground, feeding on plant material and small invertebrates. **Status** Rare and very local breeding visitor and passage migrant, presumably present late April to early September. Recorded in the breeding season in a few widely scattered river valleys in western and northern Mongolia, but nesting has never been confirmed.

Japanese Quail *Coturnix japonica*   17–19 cm

ID Adult male has rufous throat and chin, very unlike head pattern of Common Quail, but other plumages often indistinguishable in the field from that species (see Common Quail for its diagnostic characters, as well as Yellow-legged Buttonquail and Corncrake). **Voice** Male calls loud and rasping, very different from liquid three-note call of Common Quail. **Habitat** Tall dry grasslands, wet meadows and other open habitats with dense vegetation. **Behaviour** Very similar to Common Quail. An extreme skulker, more often heard than seen. **Status** Uncommon to fairly common breeding visitor and passage migrant; widespread in river valleys in north-central and eastern Mongolia. **Conservation** Considered Near Threatened globally.

Yellow-legged Buttonquail *Turnix tanki*   15–16 cm

ID Resembles *Coturnix* quails (but in fact belongs to its own family and more closely related to cranes). Hard to see, but with good views is readily distinguished from quails by richer coloration. Female is brightest with bright rufous breast and nape. Both sexes have black-spotted flanks, scaly (not streaked) back and yellow legs with three toes. Compare also rail species. **Voice** Song begins as a low-pitched hoot, becoming stronger and more drawn-out; also *off-off-off* and *pook-pook*; *hoon-hoon-hoon-hoon-hoon*; *uhuu, uhuu, uhuu* or *pwoo pwoo pwoo*. **Habitat** Grassland, scrub and marshland. **Behaviour** Shy and skulking; keeps to cover, and best located by voice. Feeds mainly on plant matter. **Status** Vagrant. One by Ulz River, Dornod province, 6 September 2002; remains of another found in Upland Buzzard's nest in valley of Nömrög River, Dornod province, 8 June 1995.

Common Pheasant *Phasianus colchicus*   M 75–89 cm, F 53–62 cm

ID Adult male unmistakable metallic rust and gold overall, with iridescent green head, large scarlet face wattle and long tail feathers. Adult female mainly brown and richly patterned, but long tail diagnostic. **Voice** Male gives loud crowing *hok-kok-kok* in display. Both sexes give similar but shriller cackle when flushed. **Habitat** River valleys with scattered willows, sandthorn (*Hippophae rhamnoides*), *Caragana* bushes and dense grasses in western Mongolia. **Behaviour** In Mongolia, very shy or skulking; will run until startled into flight. Feeds and nests on ground, but may roost in trees. **Status** Rare and local resident breeder. Two isolated populations occur in Mongolia: *P. c hagenbecki*, endemic in western Altai region; and *P. c. pallasi*, which reaches Khalkh and Nömrög River basins, Dornod province, in the east. **Conservation** Native from the Caucasus to Japan, but widely introduced as one of world's most popular game birds. However, the species is Near Threatened in Mongolia and listed in the Mongolian Red Book (2013). [Alt: Ring-necked Pheasant]

Daurian Partridge

♀

♂

Common Quail

♂

juvenile

♂

♀

♀

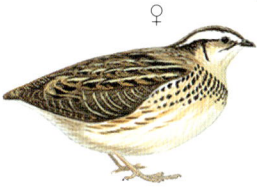

Japanese Quail

♂

Yellow-legged Buttonquail

♂

Common Pheasant

♀

Greylag Goose *Anser anser*   76–89 cm

ID Combination of overall greyish (vs. brownish) coloration, unmarked pinkish bill, absence of white on face and absence of prominent black bars on belly distinguish this species from the four other 'grey geese' that occur in Mongolia. In flight, the very pale forewings ('lags') are diagnostic. **Voice** Quite musical, with varied honking and cackling calls. **Habitat** Lakes, ponds and rivers. **Behaviour** Social. Grazes in wetlands and also feeds on water. **Status** Uncommon breeding visitor and passage migrant across Mongolia, mid-April to early October.

Greater White-fronted Goose *Anser albifrons*   64–78 cm

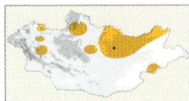

ID Adult distinguishable from other grey geese (except Lesser White-fronted, see below) by extensive white area surrounding bill (Bean Goose may have limited white edging) and by heavy black belly barring. Juv Greater White-fronted separated from Bean by distinctive bill markings. **Voice** Noisy in flocks, with quite shrill and musical honking and laughter-like *widawink widawink* calls. **Habitat** Large rivers and lakes. **Behaviour** Highly social on migration. Forages in wet meadows and agricultural land and roosts mainly on or near water. **Status** Rare passage migrant, mainly to valleys in north and east, late April to early May and late August to early September. **Conservation** Considered Vulnerable globally and Near Threatened in Mongolia due to decreasing population. Listed as Threatened in the Mongolian Red Book (2013).

Lesser White-fronted Goose *Anser erythropus*   56–66 cm

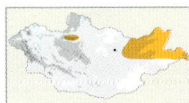

ID Smallest of the grey geese. Adult distinguished from similar Greater White-fronted Goose by shorter, daintier bill, prominent yellow eye-ring and more extensive white on face, extending to the crown. Juv similar but lacks white on face. For comparison with other grey geese, see Greater White-fronted, Greylag and Bean Geese. **Voice** Shriller and more yelping than Greater White-fronted. **Habitat** Wet grasslands, riverbanks, lakes and wheat fields. **Behaviour** Tends to occur singly or in small groups among flocks of other grey geese. Walks more quickly than larger species, often leading the way in moving flocks. **Status** Rare passage migrant, mainly in eastern Mongolia, late April to early May and late August to early September. **Conservation** Considered Vulnerable globally and Near Threatened in Mongolia. Listed as Threatened in the Mongolian Red Book (2013).

Bean Goose *Anser fabalis*   69–88 cm

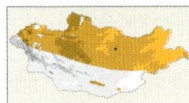

ID Adult and juv separated from other grey geese by comparatively dark brown head and neck and relatively long narrow bill with distinctive black base and tip. Lacks belly barring of white-fronted goose species and only rarely shows white around bill. **Voice** Lower-pitched, slower-paced and throatier than other grey geese. **Habitat** Lakes, ponds and other wetlands. **Behaviour** Grazes wet meadows and agricultural fields. Social. **Status** Uncommon passage migrant across Mongolia, late April to early May and late August to early September. Despite presence of suitable breeding habitat in Hövsgöl region, nesting never recorded. **Taxonomy** Two subspecies, *A. f. serrorostris* and *A. f. middendorffi*, occur in Mongolia. *A. f. serrorostris* is sometimes considered to be a separate species, Tundra Bean Goose, while *A. f. middendorffi* remains as a subspecies of Taiga Bean Goose *A. fabalis*. The status of the two forms in Mongolia is poorly known.

Great Cormorant *Phalacrocorax carbo*   80–100 cm

ID In flight or on water, may be mistaken for divers or larger grebes, but unlike the latter, adults almost wholly black with a distinctive white spot on the lower flanks. Young birds with pale underparts show a rounded, hooked (not pointed) bill on water, and in flight a long tail, unlike short tails and protruding legs of divers and grebes. On land, cormorants stand upright or perch in trees. **Voice** Throaty croaks and gurgles, especially at nest. **Habitat** Lakes and rivers with abundant fish. **Behaviour** Breeds in colonies, making stick nests on land (often islands) on cliffs or in stands of trees. Swims very low in water and dives for fish; often rests with wings outstretched. **Status** Common to locally abundant (in colonies) breeding visitor on large rivers and lakes across northern half of country and fairly common passage migrant throughout, late April to late September.

Greylag Goose

adult

juvenile

adult

adult

adult

juvenile

Greater White-fronted Goose

juvenile

Lesser White-fronted Goose

adult

middendorffi

adult

serrorostris

breeding

Great Cormorant

juvenile

adult

Bean Goose

non-breeding

Bar-headed Goose *Anser indicus*

71–76 cm

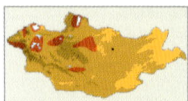

ID Unmistakable due to striking black head bars and neck pattern, golden-yellow bill and light grey plumage. **Voice** Deep, nasal *gaaah gaaah gaaah*. **Habitat** Large rivers and lakes, wet meadows and agricultural land. **Behaviour** Nests in abandoned raptor nests in trees and on cliffs. **Status** Fairly common breeding visitor and passage migrant in favourable habitat throughout Mongolia, early April to early September.

Swan Goose *Anser cygnoides*

81–94 cm

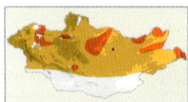

ID Black bill, elongated (swan-like) head, and sharply bicoloured head and neck are diagnostic. **Voice** Cackling and honking calls, similar to Greylag and domestic geese. **Habitat** Large lakes and rivers. **Behaviour** More closely associated with water than other grey geese, but also feeds on land. **Status** Fairly common breeding visitor and passage migrant, late April to early September. **Conservation** Considered Vulnerable globally and Near Threatened in Mongolia, though the Mongolian population is relatively high and considered stable. Listed as Threatened in the Mongolian Red Book (2013).

Mute Swan *Cygnus olor*

125–160 cm

ID Best distinguished from Whooper and Tundra Swans by colour and pattern of bill: orange with black knob in adults and greyish-pink with heavy black base in juveniles. The typically curved neck posture and longer, upturned tail of this species are also distinctive. **Voice** Low grunts and hisses when nesting; otherwise usually silent. **Habitat** Large rivers and lakes. **Behaviour** Feeds mainly on water. Very aggressive to other water birds when breeding, adopting distinctive posture with wings held high and neck tucked back when attacking. **Status** Rare and local breeding visitor and passage migrant on lakes Uvs, Bööntsagaan, Orog, Ulaan and Ögii, and Khovd and Orkhon rivers, mid-April to early September. **Conservation** This species is considered as Near Threatened in Mongolia and listed in the Mongolian Red Book (2013).

Whooper Swan *Cygnus cygnus*

140–165 cm

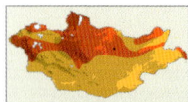

ID Extensive yellow bill base with black tip distinguishes adult from other swans. Juv shows same pattern on pinkish bill. At distance, neck typically held straighter and tail shorter than Mute Swan. **Voice** Very vocal, with loud musical bugling call frequently given in flight and at rest – typically three or four *a-hoo* notes. **Habitat** Lakes, ponds and large rivers. **Behaviour** Very social in winter and on migration. Pairs display and call together frequently. **Status** Fairly common breeding visitor and passage migrant throughout Mongolia, mid-April to mid-October. Least frequent in driest regions of south. Some winter on unfrozen rivers and lakes in western Mongolia. By far the most commonly encountered swan in the country. **Conservation** Despite relative abundance in Mongolia, listed as Very Rare in the Mongolian Red Book in 1997, partly in deference to its long-standing religious significance.

Tundra Swan *Cygnus columbianus*

120–150 cm

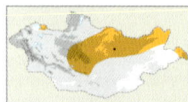

ID Smaller and with shorter neck than similar Whooper Swan, but adult best distinguished by more restricted yellow at base of bill, with a similar pattern on pinkish bill of greyish juvenile. **Voice** Similar to Whooper, but higher-pitched; calls usually single or in twos. **Habitat** Lakes and large rivers. **Behaviour** Sociable on migration. Feeds on plant material on land and in water. **Status** Uncommon passage migrant throughout much of Mongolia, late April to early May and late August to mid-September. Recorded more frequently in east. **Taxonomy** *C. c. bewickii*, sometimes considered a distinct species (Bewick's Swan), is the subspecies that occurs in Mongolia.

Bar-headed Goose

juvenile

adult

adult

adult

Swan Goose

Mute Swan

immature

adult

Whooper Swan

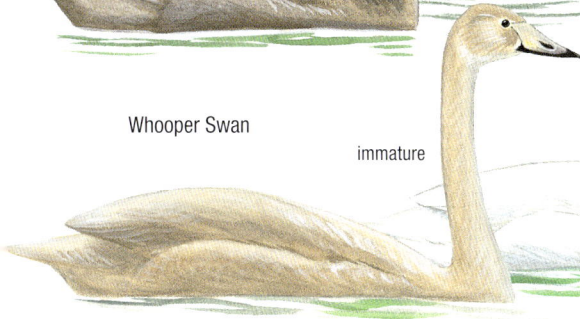

immature

adult

Tundra Swan

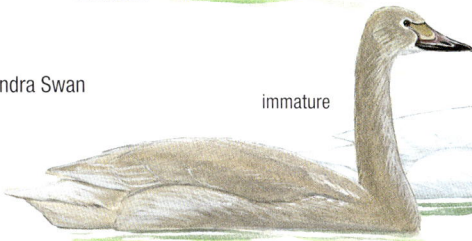

immature

adult

Red-throated Diver *Gavia stellata*   53–69 cm

ID Breeding plumage distinguished from Black-throated Diver by reddish-brown (vs. black) throat patch, and unmarked (vs. boldly spotted) back pattern. In winter plumage, paler and more slender than Black-throated with conspicuous white face. On water, holds its head and notably slender bill at upturned angle. In flight, head and neck are slung below line of body. Great Cormorant is larger and much darker overall with a stouter, hooked bill. **Voice** Largely silent on migration. Commonest calls are 'a goose-like rolling growl… a mewing wail… and a loud *kark*' (Pough 1951). **Habitat** Nests on ponds in Arctic tundra and winters at sea; on migration prefers larger freshwater lakes, but also occurs on saline lakes and large rivers. **Behaviour** Feeds on fish captured during frequent prolonged dives from the surface. **Status** Rare passage migrant, late April to early May and late August to early September; recorded only in north-central Mongolia (e.g. Buur River, Selenge province). [Alt: Red-throated Loon]

Black-throated Diver *Gavia arctica*   58–73 cm

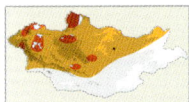

ID Bold neck and back pattern of breeding plumage unmistakable. In winter plumage, best distinguished from Red-throated Diver by stockier form, horizontal (not up-tilted) head and bill, darker overall coloration and usually a prominent white flank patch along the water line. In flight, head and neck held in line with body, not low-slung. **Voice** On breeding grounds quite vocal, with whistling *ohwiii-ka* and various croaking calls; otherwise usually silent. **Habitat** Typically nests on islands or along remote shores of large lakes; on migration, also uses large rivers; winters at sea. **Behaviour** Nest is mound of aquatic plants piled in or very close to water. Feeds on fish caught during prolonged dives from surface. **Status** Fairly common breeding visitor and passage migrant in lake districts of north, west and central Mongolia, late April to mid-September; rarely recorded in far east and south. [Alt: Arctic Loon]

Lesser Whistling Duck *Dendrocygna javanica*   41–42 cm

ID Long neck, upright posture and colour pattern of this goose-like duck is unlike that of any other Mongolian waterfowl species. Rufous colour may suggest Ruddy Shelduck but is restricted to lower breast and belly. **Voice** Flight call is a whistling *sweesik*. **Habitat** Well-vegetated lakes, ponds and wetlands. **Behaviour** Walks with ease on land. Active at dawn and dusk. **Status** Vagrant, known from single record of pair on Herlen River, Dornod province, 14 May 1998.

Ruddy Shelduck *Tadorna ferruginea*   58–70 cm

ID An unmistakable large, bright rusty-orange, goose-like duck, with prominent white patches above and below. Male has black neck-ring. **Voice** Noisy; gives various calls by night and day, e.g. goose-like *aaahng aahng* and rolling growl *ahrrrr*. **Habitat** At home in open uplands, including grain fields, as well as all types of water bodies and wetlands. **Behaviour** Nests in depressions, abandoned mammal burrows or other terrestrial cavities, sometimes far from water. May form flocks with other wildfowl in winter. **Status** Common breeding visitor and occasionally abundant passage migrant, late March to early September (late September in the Gobi). Scarce only in driest areas of Gobi. Winters locally on Lake Khar-Us, the Tuul River, wastewater ponds in Ulaanbaatar and other water bodies that remain open after autumn freeze.

Common Shelduck *Tadorna tadorna*   55–65 cm

ID Combination of dark green head, red bill and broad chestnut breast-band against largely white plumage distinguish this species from all other Mongolian waterfowl. Juv also distinctive: sooty-brown above, white below with pink bill and prominent eye-ring. **Voice** Whistling *sweees sweees* call and guttural growling on water and in flight. Female has a quacking *gagagaga* call. **Habitat** Open upland habitats as well as lakes and wetlands. **Behaviour** Nests in abandoned mammal burrows and other terrestrial cavities. Forages on land and in water. **Status** Locally common breeding visitor and uncommon passage migrant throughout Mongolia, late March to mid-September; scarcest in most arid areas of south.

Red-throated Diver

breeding

non-breeding

non-breeding

breeding

juvenile

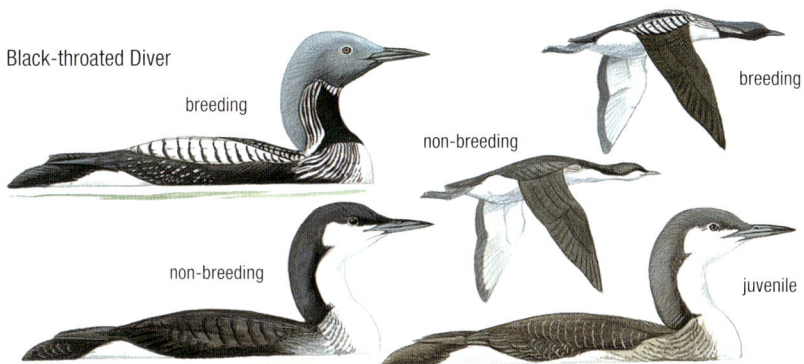

Black-throated Diver

breeding

non-breeding

breeding

non-breeding

juvenile

Lesser Whistling Duck

adult

juvenile

adult

Ruddy Shelduck

♂ breeding

♀

juvenile

♂

♀

juvenile

Common Shelduck

Mallard *Anas platyrhynchos*   50–65 cm

ID Male's metallic green head, dull yellow bill and purplish-chestnut breast separated by narrow white neck-ring are diagnostic. Female best distinguished from other drab female ducks (especially Gadwall) by orange bill with dark centre. Eclipse male like female but retains yellow bill and black rump. In flight, all plumages have metallic blue/purple speculum with white borders. **Voice** Female gives familiar loud quacking call; male has softer, more whistling *kwehp*. **Habitat** Lakes, ponds, marshes and rivers as well as fields and other open uplands, especially on migration. **Behaviour** Sociable. Feeds on plant matter, mainly on water and wet shores. **Status** Common breeding visitor and passage migrant throughout Mongolia, early April to mid-October; winters locally where there is open water, e.g. saline lakes, large rivers and water treatment ponds.

Eastern Spot-billed Duck *Anas zonorhyncha*   58–63 cm

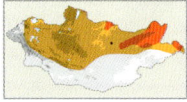

ID Largest dabbling duck, notably dark-bodied with contrasting pale head; this and prominent yellow bill tip are definitive. Sexes similar. In flight, shows prominent, white-bordered blue speculum (similar to Mallard) and white patch on rear hind corner of wing. **Voice** Mallard-like quacking, but more emphatic. **Habitat** Lakes, large ponds and rivers. **Behaviour** Sociable, especially in winter when it may flock with other dabbling ducks. **Status** Rare breeding visitor, mainly in the north-east and east of country, though encountered locally as rare passage migrant over most of country. A few individuals winter where there is open water. **Taxonomy** Formerly considered a subspecies of Indian Spot-billed Duck *A. poecilorhyncha*. [Alt: Chinese Spot-billed Duck]

Eurasian Teal *Anas crecca*   34–43 cm

ID The smallest dabbling duck. Male's chestnut and green head and prominent yellowish patch on undertail are diagnostic. Female is very similar to other female small ducks; for distinctions, see Garganey and Baikal Teal. In flight, both sexes show bright green speculum, boldly bordered with white along both edges. **Voice** Male gives a sharp rolling *pryiiik*. Female gives gruff and raspy quacks. **Habitat** Lakes, ponds, rivers and all types of wetlands. **Behaviour** Rather shy, tending to keep close to vegetated edges. Agile and wader-like in flight. **Status** Fairly common breeding visitor across well-watered northern half of country and uncommon passage migrant on lakes in desert steppe and Gobi, mid-April to early October. A few overwinter where rivers and lakes remain ice-free. [Alt: Common Teal]

Baikal Teal *Sibirionetta formosa*   39–43 cm

ID Male's bold, colourful head pattern is unique. Female distinguished from Eurasian Teal and Garganey by pale spot surrounded by dark patch at base of bill and (sometimes) pale crescent cheek patch. In flight, both sexes show bold white border along trailing edge of speculum and chestnut border along leading edge. **Voice** Male makes deep *klo klo klo* or *wo wo wo* calls, and female has a harsh, low-pitched quack. **Habitat** Like other teal species, frequents wide variety of aquatic habitats as well as agricultural fields. **Behaviour** Sociable on migration. Agile in flight. **Status** Rare passage migrant in north and east, late April to early May and early September to early October. Recorded from Orkhon, Selenge, Onon, Balj, Ulz, Khurkh and Khalkh River basins and lakes within these valleys. **Conservation** Regarded as Vulnerable globally and in Mongolia. Listed as Threatened in the Mongolian Red Book (2013). **Taxonomy** Formerly placed in genus *Anas*.

Garganey *Spatula querquedula*   37–41 cm

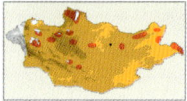

ID Small, but with longer, heavier bill than teal species. Adult male's dark brown head and breast with bold white stripe over eye are diagnostic. Adult female distinguished from teal species by prominent facial stripes and bill shape. In flight, wings of male are largely pale grey above, and both sexes have speculum with white borders on both edges. **Voice** Male gives a hard, wooden croaking *kaarrrr*; female has a soft quack. **Habitat** Occurs in wide range of aquatic habitats, but prefers shallow, well-vegetated waters and shores; also feeds in grain fields. **Behaviour** Quite shy, especially when nesting. Social away from breeding grounds. **Status** Uncommon and local breeding visitor across well-watered northern regions and uncommon passage migrant in desert steppe and Gobi, mid-April to early October. **Taxonomy** Formerly placed in genus *Anas*.

♂

♀

♀

♂

Mallard

Eastern Spot-billed Duck

♂

♂

♀

Eurasian Teal

♂

♀

Baikal Teal

♀

♀

♂

♀

♂

Garganey

Falcated Duck *Mareca falcata*   46–54 cm

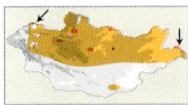

ID Bottle-green head with dark chestnut forehead and crown and distinctive head shape with nape crest or 'mane' distinguish male from all other waterfowl. Female best told from other female dabbling ducks by notably plain greyish, rather square head and blackish bill. In flight, both sexes show wing pattern similar to Garganey, but lack the bold white trailing edge of the speculum. Female distinguished from female Eurasian Wigeon in flight by lack of white belly. **Voice** Male has a two-note *foo-wee* call, sometimes followed by a buzzing *brurururu*. Female has a hoarse quack. **Habitat** In Mongolia, may prefer forest and forested steppe lakes and swamps for breeding but appears in other aquatic habitats on migration. **Behaviour** Associates with other dabbling ducks outside of breeding season. **Status** Rare and local breeding visitor across northern Mongolia and rare passage migrant throughout, mid-April to mid-September. **Conservation** Considered as Near Threatened and decreasing globally and in Mongolia. Listed as Near Threatened in the Mongolian Red Book (2013). **Taxonomy** Formerly placed in genus *Anas*. [Alt: Falcated Teal]

Gadwall *Mareca strepera*   46–58 cm

ID Male is a rather uniformly drab grey duck with black rear end, quite unlike other male dabbling ducks. Female closely resembles female Mallard, but slightly smaller with a shorter, narrower bill; orange bill has even, blackish centre line (vs. Mallard, in which black is blotchy). In flight, both sexes have white (not buffy) belly and prominent white speculum (also visible at rest). **Voice** Male gives whistling and croaking calls; female quacks like Mallard but with higher, more nasal sound. **Habitat** Virtually all aquatic habitats but favours shallow, well-vegetated still waters. **Behaviour** Shy on breeding grounds; flocks with other dabbling ducks at other times. **Status** Common breeding visitor and fairly common passage migrant throughout Mongolia, including desert steppe and Gobi on migration, mid-April to early October. **Taxonomy** Formerly placed in genus *Anas*.

Eurasian Wigeon *Mareca penelope*   45–51 cm

ID Male has yellowish-buff forehead and pinkish breast. Female is fairly uniform rufous-brown, with small, rounded head compared to other female dabbling ducks and pale blue-grey bill. In flight, male shows prominent white forewing patch, and both sexes are white-bellied. **Voice** Male has distinctive whistled *wheeoiio* call. **Habitat** Wide variety of aquatic habitats, favouring well-vegetated shallow bodies when breeding, but appearing in ponds, desert lakes, wheat fields etc. on migration. **Behaviour** Grazes on land more often than most other dabbling ducks. Very sociable in winter and on migration. **Status** Fairly common breeding visitor, mainly in Lake Uvs, Great Lakes Depression and Hövsgöl region, and uncommon passage migrant elsewhere, including desert steppes and Gobi, mid-April to early October. **Taxonomy** Formerly placed in genus *Anas*.

Northern Pintail *Anas acuta*   50–65 cm

ID Large, elegant dabbling duck with long neck and long, pointed tail. Male's chocolate-brown head and nape, with white vertical neck stripe contrasting with pale grey and white elsewhere, and very long 'pin' tail are unmistakable. Female distinguished from other dabbling ducks by slender, elongate proportions. In flight, long neck and tail and slender form are best indicators for both sexes. **Voice** Generally quiet. Male gives various soft whistles and growls; female has a hoarse quack. **Habitat** Like other dabbling ducks, favours shallow, well-vegetated waters, especially when breeding, but occurs in all manner of water bodies and wetlands as well as grain fields during migration. **Behaviour** Flocks with other dabbling ducks away from breeding grounds. **Status** Fairly common breeding visitor in the Lake Uvs and Great Lakes Depression and uncommon passage migrant throughout, mid-April to early October.

Northern Shoveler *Spatula clypeata*   43–56 cm

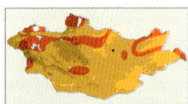

ID Adult male's combination of green head, chestnut flanks, white breast and notably large bill unmistakable. Female best distinguished by bill, but often feeds with head submerged for long stretches, when it can resemble other dabbling ducks, especially female Mallard. In flight, bold pattern of male includes blue forewing patch, which is grey in female. **Voice** Quiet. Male may give rattling call when flushed; female has a soft quack. **Habitat** Prefers still, shallow, well-vegetated waters when breeding, but uses a wide range of other habitats on migration, including wet grain fields. **Behaviour** Feeds in shallows, immersing bill to dabble for long spells, upending less often than other dabblers. Sociable in non-breeding period. **Status** Fairly common breeding visitor and passage migrant throughout Mongolia apart from driest regions of far south, mid-April to early October. **Taxonomy** Formerly placed in genus *Anas*.

♂
♀

Gadwall

♂

♂
♀

Falcated Duck

♂
♀

♂
♀

♂

Eurasian Wigeon

♀

♂
♀

Northern Pintail

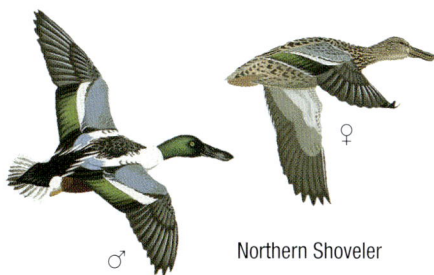

♂

♂
♀

Northern Shoveler

♂

♀

♂

♀

Mandarin Duck *Aix galericulata* 41–51 cm

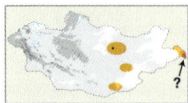

ID Smallish duck with dainty proportions. Adult male is unmistakably ornate, with reddish bill, orange 'mane', upstanding wing 'sails' and multiple colours and patterns. Female best told by distinctive head shape, white 'spectacles' and pale spotting on breast and flanks. In flight, shows dark wings with white trailing edges and white belly. **Voice** Male gives various pleasant whistling calls; female gives deeper clucking notes. **Habitat** Breeds on secluded forested rivers and streams. On migration, may appear at almost any water body usually with well-vegetated margins. **Behaviour** Nests in tree cavities and often perches and roosts in trees. **Status** Possibly a rare and very local breeding visitor in Nömrög River valley and Khyangan region of far east; a very rare passage migrant in some areas of central and southern Mongolia, mid-April to late September.

Red-crested Pochard *Netta rufina* 53–58 cm

ID Male's puffy, golden-orange head, bright red bill and parti-coloured plumage are unmistakable. Female best told from *Aythya* pochards by pale grey cheeks and sharply contrasting dark crown; also by rounded head and long pointed bill. In flight, male's pied plumage stands out; white, black-bordered wings with pale grey body distinguish female. **Voice** Rasping and growling calls given in courtship, higher-pitched in male. **Habitat** Most typically, large lakes with well-vegetated shorelines, but also other shallow bodies. **Behaviour** Nests along lakes and surrounding wetlands. Feeds by dabbling and upending as well as diving. **Status** Uncommon breeding visitor, mainly in western Mongolia, and fairly common passage migrant elsewhere, mid-April to late September. **Taxonomy** Formerly placed in genus *Rhodonessa*.

Common Pochard *Aythya ferina* 42–58 cm

ID Adult male's chestnut head with uniform grey back and flanks, and black breast and rear, are distinctive. Female distinguished from other *Aythya* females by paler coloration overall, sloping forehead, 'spectacles' and (usually) prominent bluish bill-band. In flight, both sexes paler overall than other pochards, with less contrast in wing pattern. **Voice** Male gives soft chattering and whistling calls; female has harsh growl, often given in flight. **Habitat** Mainly lakes with extensive open water, but also other water bodies, especially on migration. **Behaviour** Feeds mainly by diving for plant matter as well as molluscs, aquatic insects; sometimes upends in shallows like a dabbling duck. Forms (often large) rafts with other *Aythya* ducks when not breeding. **Status** Fairly common breeding visitor, especially in western and central lakes regions, and fairly common passage migrant elsewhere, mid-April to early October. **Conservation** Considered Vulnerable globally.

Ferruginous Duck *Aythya nyroca* 38–42 cm

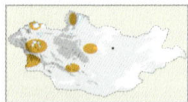

ID Smallest *Aythya* species, with a notably peaked head and rich, very dark chestnut head, breast and flanks contrasting with bright white undertail in both sexes, though female duller. Male has prominent whitish eye. Most confusable with female Tufted Duck but the latter is dull brown with flatter head and a yellow eye. Baer's Pochard is also duller and has white flanks in all plumages; see also distribution in Mongolia. In flight, shows extensive white wing-stripe and white belly. **Voice** Generally silent, though in courtship male gives whistles and nasal cackle; female gives low-pitched snorting *err err err* or *kakaka*. **Habitat** Shallow lakes with extensive emergent vegetation as well as river shallows and marshes. **Behaviour** Shy. Feeds mainly on aquatic vegetation by dabbling and upending as well as diving. **Status** Rare passage migrant to lakes in western and central Mongolia, late April to early May and late August to early September. **Conservation** Considered Near Threatened globally and Vulnerable in Mongolia. Listed as Threatened in the Mongolian Red Book (2013). [Alt: Ferruginous Pochard, White-eyed Pochard]

Baer's Pochard *Aythya baeri* 41–46 cm

ID A very rare and local pochard, best distinguished in favourable conditions by partially white flanks. Lacks rich chestnut colour of Ferruginous Duck (which does not co-occur in Mongolia). Male Tufted Duck has black (vs. brown) breast and clean cut, all-white flanks and crested head. Female Tufted lacks white on flanks. Identifiable in flight only with great caution. **Voice** Silent except during courtship, when gives harsh growling notes. **Habitat** Most typically, shallow freshwater lakes with extensive emergent vegetation but also uses rivers, ponds and other water bodies on migration. **Status** Rare and local passage migrant, late April to early May and late August to early September. Recorded on Lakes Buir and Tashgain Tavan, and Khalkh and Nömrög Rivers; also small lakes of Khyangan in far eastern Mongolia. Paired birds have been observed but breeding not established. **Behaviour** Shy, but flocks with other *Aythya* ducks on migration. **Conservation** Considered as Critically Endangered globally and listed in the Threatened Birds of Asia (2001).

Mandarin Duck

Red-crested Pochard

Common Pochard

Ferruginous Duck

Baer's Pochard

Tufted Duck *Aythya fuligula*   40–47 cm

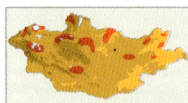

ID Male distinguished from Greater Scaup by black (vs. grey) back and at least some trace of crest on back of head. Female sometimes has prominent white area around base of bill (like Greater Scaup), but usually has some hint of crest. Female may also have whitish undertail-coverts and might be confused with Ferruginous Duck and Baer's Pochard, which see for distinctions. **Voice** Generally silent, but male gives whistling calls in courtship. Female has typical *Aythya* duck *krrr krrr* growl, also given by male on take-off. **Habitat** Typically freshwater and saline lakes with well-vegetated margins, but also occurs on calm areas of large rivers and other waters, especially on migration. **Behaviour** Feeds mainly by diving for molluscs, aquatic insects and some plant matter. Very social with other pochards away from breeding grounds. **Status** Common breeding visitor and passage migrant in appropriate habitats, including lakes and wetlands of desert steppe and Gobi, mid-April to early October. Winters on open waters of Lake Khar-Us and Tuul River near Ulaanbaatar city.

Greater Scaup *Aythya marila*   40–51 cm

ID Male distinguished from smaller Tufted Duck by grey (vs. black) back and absence of crest. Female typically shows prominent clear white area surrounding bill, which female Tufted usually lacks (or shows just a very small patch). Tufted also tends to show modest crest. In flight, paler back and forewings above, as well as heavier build, are the best pointers. See also Baer's Pochard and Ferruginous Duck. **Voice** Usually silent; courting drakes said to give 'soft cooing and whistling notes' (Madge and Burn 1988). **Habitat** Prefers freshwater and saline lakes, but can be present on any water bodies where congregations of migratory waterfowl occur. **Behaviour** Dives for aquatic invertebrates and vegetation. Flocks with other diving ducks in winter and on migration. **Status** Vagrant. One on Lake Uvs, Uvs province, June 1964, one on Lake Ögii, Arkhangai province, June 2006 and two on Lake Orog, Bayankhongor province, July 2006.

Harlequin Duck *Histrionicus histrionicus*   38–45 cm

ID Very distinctive, small, compact, round-headed duck with tiny bill. Male's slate-blue and chestnut plumage with 'clown' face unmistakable. Female also distinctive: uniform chocolate-brown with cleanly contrasting facial spots; however, compare White-winged Scoter. **Voice** Highly vocal with piping calls, though normally silent on migration. **Habitat** Breeds near fast-flowing boreal rivers and winters on rocky sea coasts. On migration may visit lakes or fast-flowing rivers. Not sociable with other diving ducks. **Behaviour** Feeds in collective flocks, mainly on aquatic and marine invertebrates. **Status** Vagrant. A single bird on Lake Orog, Bayankhongor province, 1–2 June 1962.

Long-tailed Duck *Clangula hyemalis*   M 51–60, F 37–47 cm

ID In winter plumage, might be confused with Smew, but always distinguishable by facial pattern and long tail of drake. Breeding plumage head pattern is also diagnostic in both sexes. **Voice** Noisy in late winter, male with low *ow-ow-owna-ow* call, female with a high quack or *kak kak kak*. **Habitat** Nests on Arctic tundra and winters at sea, but on migration may visit lakes or other waters. **Behaviour** Fast and graceful in flight; makes deep dives mainly for molluscs, crustaceans and small fish. **Status** Vagrant. One on Lake Orog, Bayankhongor province, 8 June 2003; pair on Lake Tashgain Tavan, Dornod province, 10 May 2004; and one on Lake Delgertsagaan, Dornod province, August 2004.

Common Goldeneye *Bucephala clangula*   42–50 cm

ID Male's bottle-green, oddly rounded head with circular spot at base of bill is diagnostic. Female superficially resembles female pochards, but note head shape, short bill, and white 'check' in wing. In flight, large dark head, contrasting with paler body, is distinctive in both sexes. **Voice** Fairly quiet. Male has squeaking and grating calls, female a low pochard-like growl. Wings make striking whistling sound in flight. **Habitat** Breeds near rivers and lakes in forested areas, but uses other water bodies on migration. **Behaviour** Nests in tree cavities, including old Black Woodpecker nest-holes. Dives for aquatic invertebrates and plant matter. **Status** Fairly common breeding visitor in forested areas of northern Mongolia, and fairly common passage migrant elsewhere, including desert steppe and Gobi, mid-April to early October. Rare winter visitor on rivers with open water in west and central Mongolia.

with scaup-like head

♀

♂

imm

♂

Tufted Duck

imm

♀

♂

imm

Greater Scaup

♂

♀

Harlequin Duck

♀

♂

♂ winter

♀ winter

♂ summer

Long-tailed Duck

♂ winter

Common Goldeneye

♂

♀

♂

♀

♀ winter

White-winged Scoter *Melanitta deglandi*   51–58 cm

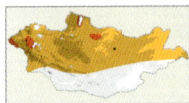

ID Adult male's large size, all-black plumage with white wing patch, extensively yellow-orange bill and white 'tear drop' under the eye are unlike any other Mongolian duck. Female somewhat resembles female Harlequin Duck, but much larger with bulkier bill and different arrangement of facial spots. In flight, dark plumage with prominent white patch on trailing edge of wing is diagnostic in both sexes. **Voice** Quiet except when displaying, when male has whistling call and female gives deep *kraa-ah* call. **Habitat** Nests near forest lakes and large rivers, but occurs in other aquatic habitats on migration. **Behaviour** Feeds by prolonged deep dives for aquatic invertebrates and some plant matter. Nests on ground in dense cover near water. **Status** Uncommon breeding visitor in northern and western Mongolia and uncommon passage migrant elsewhere, including lakes in dry steppe and Gobi, mid-April to late September. Typically winters in flocks at sea but may linger on inland waters that remain open. **Taxonomy** Previously considered conspecific with Velvet Scoter *M. fusca* of Europe and Western Asia. Mongolian birds are now considered to be one of two subspecies of White-winged Scoter *M. d. stejnegeri*, sometimes called Siberian Scoter.

White-headed Duck *Oxyura leucocephala*   43–48 cm

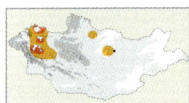

ID Unmistakable. Adult male differs from all other ducks by white head, 'swollen' blue bill and chestnut breast. Female distinguished by large grey bill, bold face stripe across pale cheek and long, stiff, often cocked tail. **Voice** Normally silent; displaying males may give 'low rattling noise(s) and piping calls' (Madge and Burn 1988). **Habitat** Large freshwater lakes and brackish lagoons with large reedbeds. **Behaviour** Dives and dabbles, mainly for plant material. Shy, often staying close to vegetated margins. Seldom flies. Sometimes adopts abandoned nests of coots or other duck species for nesting. **Status** Rare and very local breeding visitor on some lakes in Great Lakes Depression and rare passage migrant in suitable habitat elsewhere, mid-April to mid-September. **Conservation** Considered as Endangered globally and in Mongolia. Listed as Threatened in the Mongolian Red Book (2013) and included in the Threatened Birds of Asia (2001).

Smew *Mergellus albellus*   35–44 cm

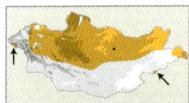

ID Small size, largely white plumage, peaked crest and 'panda face' of male unmistakable. Female faintly resembles small female Common Goldeneye, but note white throat and slender bill. Compare also Long-tailed and White-headed Ducks. **Voice** Quiet away from breeding grounds. Courting male has croaking call, female a growling or cackling *krrr*. **Habitat** Large lakes and rivers. **Behaviour** Dives at length for fish, aquatic invertebrates and plant matter. Quite sociable away from breeding grounds, especially with Common Goldeneye. **Status** Rare and local passage migrant, late April to early May and late August to late September. Irregular and very local winter visitor on some ice-free lakes and rivers in Great Lakes Depression.

Red-breasted Merganser *Mergus serrator*   52–58 cm

ID Male distinguished from similar Goosander by smaller size, slimmer form, long 'hairy' crest and reddish-brown, streaked (as opposed to clear whitish) breast. See similar female Goosander for distinctions. In flight, both sexes present generally darker appearance than Goosander, with more black striping in white wing patches. **Voice** Assorted hiccupping, sneezing and purring calls in courtship. **Habitat** Lakes and large rivers with abundant fish. **Behaviour** Nests near water on ground or in tree cavities. Dives frequently, feeding mainly on fish. Highly social outside breeding season. **Status** Possibly very rare and local breeding visitor, but not confirmed; rare passage migrant in north-central Mongolia, late April to early May and late August to early September.

Goosander *Mergus merganser*   58–66 cm

ID Adult male separated from Red-breasted Merganser by large size, clear whitish breast and flanks, and smooth (not ragged) nape crest. Female has deeper rufous head and upper neck than female Red-breasted Merganser, sharply divided from grey breast and clean-cut white throat patch. In flight, both sexes show more white in wings than Red-breasted Merganser. **Voice** A rolling or rattling *karoo-kraa* or *karrr karrr*. **Habitat** Deep, freshwater lakes and fast-flowing rivers and streams. **Behaviour** Nests in cavities in mature trees in forested areas near water. Catches fish in prolonged dives. Sociable outside breeding season. **Status** Common breeding visitor in large rivers in north and north-east Mongolia and fairly common passage migrant in appropriate habitats elsewhere, mid-April to early October. Rare and local winter visitor on rivers that remain ice-free. [Alt: Common Merganser]

White-winged Scoter

♂ ♀ ♀ ♂

White-headed Duck

♀ ♂ ♀ ♂

Smew

♂ ♀

Red-breasted Merganser

♂ ♀ ♂ ♀

Goosander

♂ ♀ ♂ ♀

Black-necked Grebe *Podiceps nigricollis*
28–34 cm

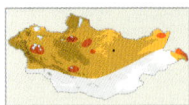

ID May be confused with Horned Grebe, from which best distinguished in breeding plumage by black, not reddish, neck and form and location of golden ear tufts. In non-breeding plumage, Black-necked has distinctive black 'sideburn', peak at top (not rear) of crown, and shorter, thinner, upturned bill. **Voice** Gives low chattering trill, plaintive *pew-ee* and firm whistling *feweeech* on territory; otherwise typically silent. **Habitat** Nests along vegetated shores of freshwater lakes and ponds; more widespread during migration. **Behaviour** Often nests in loose colonies; nest is mound of aquatic vegetation anchored to floating weed mats or built up in shallow water. Dives frequently for small aquatic insects and crustaceans. **Status** Locally fairly common breeding visitor in central Mongolia, from Khovd province in west to eastern Dornod, and common passage migrant on fresh water elsewhere, late April to early September. [Alt: Eared Grebe]

Horned Grebe *Podiceps auritus*
31–38 cm

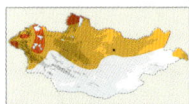

ID Likely to be confused only with Black-necked Grebe. In breeding plumage easily distinguished from that species by reddish-brown neck and location of golden ear tufts. In non-breeding plumage, white of cheeks is more extensive; head is flatter with peak at rear; bill is straighter and heavier. **Voice** On breeding grounds gives various throaty rattling calls and a shrill, breathy trill; otherwise mainly silent. **Habitat** Nests along vegetated shores of wide variety of freshwater bodies; even more widespread during migration. **Behaviour** Often breeds in loose colonies; nest is mound of water plants either floating and attached to emergent vegetation, in shallow water, or on dry land next to shore. Dives from surface for aquatic insects and crustaceans. **Status** Locally common breeding visitor on Lake Hövsgöl and in Altai and common passage migrant on fresh water elsewhere, late April to early September. **Conservation** Considered Vulnerable globally. [Alt: Slavonian Grebe]

Red-necked Grebe *Podiceps grisegena*
40–50 cm

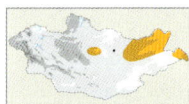

ID A large grebe, though smaller and notably stockier than Great Crested. Reddish-brown neck and breast, white cheeks and yellow bill base of breeding plumage unmistakable. In non-breeding plumage distinguished from Great Crested Grebe by dark face, shorter, thicker neck and black-tipped bill with yellow base. In flight, shows white bars on both leading and trailing edges of wings, but more restricted than in Great Crested. **Voice** On breeding territory gives various gull-like wails, squeals and whinnying trills; pairs give low *uwahh uwahh* calls together; otherwise mainly silent. **Habitat** Nests in vegetated margins of lakes, ponds and slow-flowing rivers. Migrants may occur in open water of all types. **Behaviour** Nest is mound of aquatic vegetation either floating or in shallow water. Dives frequently for fish, crustaceans and aquatic insects. **Status** Very rare passage migrant, late April to early September. In Mongolia, migrating birds known only from north-east in Hentii and Dornod provinces.

Great Crested Grebe *Podiceps cristatus*
45–52 cm

ID Unmistakable in breeding plumage due to large size, slender build, white neck and striking head plumes. In non-breeding plumage might be mistaken for larger, heavier Black-throated or Red-throated Divers, but note brownish flanks and dark line from eye to bill. In flight, divers lack white wing patches. Told from smaller, chunkier Red-necked Grebe by longer, more slender neck and paler coloration, especially of head and neck. **Voice** Various harsh croaking and rolling calls and steady low *breck breck breck* between pairs when nesting; otherwise mainly silent. **Habitat** Well-vegetated lake and pond shores while nesting and wider range of water bodies on migration. **Behaviour** Nest is pile of aquatic vegetation anchored to emergent plants or piled up in shallow water. Pairs perform elaborate courtship display on surface of water. Dives from surface, mainly for fish but also aquatic insects and crustaceans. **Status** Locally common breeding visitor (except southern Gobi) and common passage migrant throughout Mongolia, late April to early September. **Conservation** Large, widespread transcontinental population appears stable.

Little Grebe *Tachybaptus ruficollis*
25–29 cm

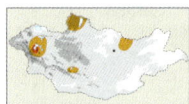

ID Smallest Eurasian grebe with 'powder puff' rear end. In breeding plumage lacks yellow ear tufts of Slavonian and Black-necked Grebes and shows a conspicuous yellow gape spot at base of bill. Non-breeding plumage is buffy, not whitish, on neck and sides with contrasting blackish crown and back. Bill is pale and stubby. In flight, most populations show restricted white on secondaries. **Voice** Loud, shrill, whinnying trills on breeding grounds; otherwise mainly silent. **Habitat** Wide variety of shallow lakes and ponds with reedbeds. **Behaviour** Not colonial. Dives frequently for aquatic insects, crustaceans, molluscs and occasionally small fish. Often rather shy, remaining close to vegetated shore edges. **Status** Rare summer visitor; may nest on Lake Khar-Us in Great Lakes Depression. Recorded on migration on other freshwater and saline lakes, including Lakes Hövsgöl, Hövsgöl province and Bööntsagaan, Bayankhongor province, late April to early September.

breeding

non-breeding

juvenile

non-breeding

Black-necked Grebe

breeding

non-breeding

juvenile

non-breeding

Horned Grebe

breeding

juvenile

on-breeding

non-breeding

Red-necked Grebe

breeding

non-breeding

non-breeding

juvenile

Great Crested Grebe

breeding

non-breeding

juvenile

breeding

Little Grebe

Black-headed Ibis *Threskiornis melanocephalus* 65–76 cm

ID White plumage with contrasting black head, neck and tail, and strongly decurved bill are unique in Mongolia. **Habitat** Wetlands and cultivated areas. **Behaviour** Forages like a stork, walking purposefully and pausing to probe ground for prey. **Status** Vagrant. Known only from a single unconfirmed report from Lake Khar-Us in Great Lakes Depression in September 2006.

Eurasian Spoonbill *Platalea leucorodia* 70–95 cm

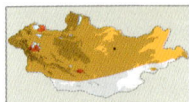

ID Possibly confusable with Great Egret at distance or when roosting with bill hidden, but has longer, stouter bill and flies with neck outstretched like a stork. Even at moderate distance, spatulate bill and yellowish nape crest are conspicuous and definitive. **Voice** Mostly silent but gives various low guttural and grumbling sounds at nest. **Habitat** Marshes, lakes and large rivers. **Behaviour** Breeds and roosts in dense colonies, constructing platforms of sticks and other plants on ground in reedbeds or in wetland bushes and trees, often on islands. Feeds by sweeping bill through water as it wades, to 'sieve' out small invertebrate prey. **Status** Very local breeding visitor at Lake Uvs and delta of Tes and Torkholig Rivers, Uvs province, and Lakes Bayan, Khar and Khar-Us in Great Lakes Depression; formerly nested at Lake Orog in Valley of the Lakes. Also an uncommon but regular passage migrant over much of country, late April to late August.

Oriental Stork *Ciconia boyciana* 110–115 cm

ID Black bill and all-black flight feathers distinctive. Might be mistaken for Siberian Crane, though latter has red face patch and bill base with white, not black secondaries. Juv Siberian Crane has rusty head and neck, unlike any stork plumage. **Habitat** Wetlands, lakes, river valleys and agricultural areas. **Behaviour** Hunts prey on foot with deliberate strides. Often soars. **Status** Very rare passage migrant and summer visitor with only a handful of records from Dornod province. **Conservation** Globally Endangered.

Black Stork *Ciconia nigra* 95–100 cm

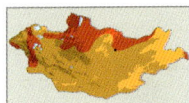

ID No other large black bird with long red legs and bill and white underparts occurs in Mongolia. Juv lacks red legs and bill, but remains unmistakable. **Habitat** River valleys, lakes and pools with fish and amphibians near forest and rocky mountains. **Voice** Said to make rasping and mewing sounds at nest, but rarely heard. **Behaviour** Breeds non-colonially, making bulky stick nests on cliffs or tall trees, especially along wooded rivers. Forages in all types of wetlands and uplands, walking in a deliberate manner as it searches for small vertebrate and large invertebrate prey. **Status** Locally fairly common breeding visitor in northern and central Mongolia and uncommon passage migrant elsewhere, mid-April to early October.

Greater Flamingo *Phoenicopterus roseus* 125–145 cm

ID Unmistakable; no other pinkish, long-legged wading bird with downcurved bill occurs in Mongolia. **Habitat** Typically, shallow saline lakes. **Behaviour** Habitually occurs in flocks. Feeds with head upside-down in water, filtering out tiny aquatic prey through the bill. **Status** Vagrant. A small flock was recorded in Lake Khar-Us, Khovd province, in 1947 (exact date unknown) and three individuals were in Shishkhid River in Darkhad depression, Hövsgöl province, in 1996 (exact date unknown).

Dalmatian Pelican *Pelecanus crispus* 160–180 cm

ID Cannot be mistaken for any regularly occurring Mongolian species, given its large size, distinctive form and massive bill. However, similar Great White Pelican *P. onocrotalus* breeds in Kazakhstan and Kyrgyzstan and could conceivably occur in Mongolia. In flight, adult Great White shows contrasting black flight feathers below, compared to relatively uniform pale grey in Dalmatian and black, not dark grey, primaries above. At close range, Great White shows a patch of pink skin surrounding the eyes, absent in Dalmatian. In full breeding condition, Dalmatian has a reddish throat pouch compared to yellow in Great White. **Voice** Gives low growling or roaring sounds at the nest; otherwise usually silent. **Habitat** Nests on large lakes; feeds in rivers and other water bodies on migration. **Behaviour** Nests colonially, sometimes with Great Cormorants, on floating islands of vegetation and other materials. Feeds almost exclusively on fish. **Status** Rare and local breeding visitor and passage migrant, late April to early September. Has bred in recent past at lakes Khar-Us and Airag (western Mongolia), and a few birds continue to attempt to nest in these same areas, though numbers much reduced in recent decades. Migrating birds have also been recorded – at times in numbers – elsewhere in the Great Lakes Depression and Valley of the Lakes. **Conservation** Considered Vulnerable globally and Critically Endangered in Mongolia. Listed as Threatened in the Mongolian Red Book (2013). Contributing to its decline in Mongolia is the long-standing tradition of using the upper mandible of pelicans as a racehorse scraper, the value of which increases, of course, as the birds become rarer.

immature

adult

Black-headed Ibis

breeding

Eurasian Spoonbill

juvenile

breeding

adult

adult

Oriental Stork

juvenile

adult

Black Stork

adult

juvenile

juvenile

immature

Greater Flamingo

adult

juvenile

breeding

adult

juvenile

Dalmatian Pelican

Eurasian Bittern *Botaurus stellaris*   64–80 cm

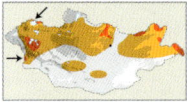

ID Most likely to be mistaken for immature Black-crowned Night Heron, but lacks white speckling on mantle, and in flight shows markedly barred flight feathers with feet extending entirely beyond tail. Brown juv Purple Heron is much more slender. **Voice** In breeding season, male gives very low-pitched, very far-carrying booming call, likened to foghorn or lowing of bull. **Habitat** Dense reedbeds near rivers and lakes. **Behaviour** Shy and skulking; best located near dawn and dusk by distinctive call. Slowly stalks fish and other vertebrate prey in dense reeds, often standing motionless for concealment. **Status** Secretive and uncommon but widespread breeding visitor in appropriate habitats and uncommon passage migrant in desert lakes and wetlands, late April to early September. **Conservation** Least Concern globally, but considered at risk in Mongolia due to decreasing numbers and widespread threats to wetland habitats. Listed as Near Threatened in the Mongolian Red Book (2013). [Alt: Great Bittern]

Little Bittern *Ixobrychus minutus*   27–36 cm

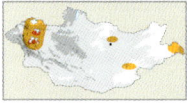

ID Very small size and prominent pale buffy wing patches are unique among Mongolian herons, except for vagrant Schrenk's Bittern. However, adult male of latter has dark chestnut neck and back (vs. tawny neck and black back of Little), and Schrenk's wing patch is darker, not whitish buff. Streaky juv Little Bittern distinguished from juv Schrenk's Heron by pale (vs. dark) brown coloration. **Voice** When nesting, a regularly repeated croaking note given mainly at night; in flight a repeated nasal *kek*. **Habitat** Dense reedbeds on lake and pond shores and river backwaters. **Behaviour** Shy and difficult to see except when flying; climbs among the reeds with agility. **Status** Rare and very local breeding visitor in reedbeds in Great Lakes Depression and possibly in suitable wetlands in Dornod province, and elsewhere very rare passage migrant, late April to early September. **Conservation** Least Concern globally, but considered at risk in Mongolia due to decreasing numbers and widespread threats to wetland habitats. Listed as Near Threatened in the Mongolian Red Book (2013).

Schrenck's Bittern *Ixobrychus eurhythmus*   33–39 cm

ID A dark chestnut-brown version of Little Bittern (see above for detailed distinctions). Otherwise, its tiny proportions and contrasting wing patches of adults make it unlike any other Mongolian heron. **Habitat** Breeds in reedbeds on lake and pond shores and riverine coves; may occur in other wetlands on migration. **Behaviour** As Little Bittern: reclusive, nesting and feeding in dense marshland vegetation. **Status** Vagrant. Only two documented records in Mongolia, from Dornod province in the 1990s: one in Mongol Daguur Strictly Protected Area, Dornod province, in June 1995 and a male at Herlen River, near Choibalsan town, 9 June 1998.

Black-crowned Night Heron *Nycticorax nycticorax*   56–65 cm

ID Stocky build, black back and bill, and white underparts distinguish adult from other Mongolian herons. Juv resembles Great Bittern but has conspicuous white speckling on mantle. Lacks Great Bittern's strongly barred flight feathers, and legs do not extend fully beyond tail in flight. **Voice** Distinctive barking *quok* given in flight, especially at dawn and dusk or when flushed. **Habitat** Lake and pond shores, reedbeds and other wetlands. **Behaviour** Roosts by day, often in trees near water, heading out to feed at dusk or under thick cloud cover. **Status** Vagrant. One at Lake Khar, Khovd province, 29 May 1995; then seven further singles from 2004, all between May and early August.

Chinese Pond Heron *Ardeola bacchus*   42–52 cm

ID Structure like Eastern Cattle Egret (see Plate 15), but with rich chestnut head, neck and breast and charcoal-grey mantle. Non-breeding and young birds are duller grey-brown and streaky. In flight, reveals pure white wings, rump and tail. **Habitat** A wide variety of open habitats, including wetlands and cultivated areas. **Behaviour** Forages in the open; not skulking. **Status** Rare, but increasingly regular passage migrant in Mongolia, with records in scattered localities, May to October.

adult

juvenile

Eurasian Bittern

♂
♀
juvenile

♂
♀

Little Bittern

juvenile

♂
breeding

♀

juvenile

♂

Schrenck's Bittern

adult

juvenile

juvenile

juvenile

adult

breeding

Chinese Pond Heron

non-breeding

Black-crowned Night Heron

Eastern Cattle Egret *Bubulcus coromandus*

46–56 cm

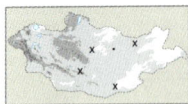

ID Differs from Great Egret by short legs and bill, stocky build and preference for upland rather than wetland habitats, and from Little Egret by yellow (vs. black) bill and relatively pale legs. In breeding plumage, conspicuous buffy patches on head, breast and mantle, and pinkish legs. In non-breeding plumage, all white with dark legs. **Habitat** Associated with open fields and shallow, open wetlands, rather than lakes and dense marshes favoured by most herons. **Behaviour** Often in flocks, following livestock or wild ungulates to feed on flushed insects. **Status** Vagrant. Known from only four records in scattered localities between 2000 and 2010, all in July. **Taxonomy** Formerly treated as conspecific with Western Cattle Egret *B. ibis*, using the name Cattle Egret.

Great Egret *Ardea alba*

80–104 cm

ID Much larger than other white herons in the region with long, yellow bill (turning black with yellow base during courtship) and long black legs and feet, which extend beyond tail in flight. Acquires long, delicate plumes on lower back during courtship. **Voice** Hoarse calls (*graaah*) when taking off and at nest site. **Habitat** Many types of wetland, especially large lakes and rivers. Usually associated with reedbeds. **Behaviour** Nests in loose colonies in reedbeds or bushes and trees where available. Not shy. **Status** Uncommon breeding visitor and passage migrant, late April to early September. Known to nest only by lakes and rivers of the Great Lakes Depression in western Mongolia; more widespread as a migrant, occurring regularly in the wetlands of eastern Mongolia and Valley of the Lakes. **Taxonomy** Formerly placed in genus *Casmerodius* or *Egretta*. [Alt: Great White Egret]

Little Egret *Egretta garzetta*

55–65 cm

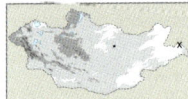

ID Much smaller than Great Egret; taller and more slender than Eastern Cattle Egret. Slender black bill and black legs contrasting with yellow feet are diagnostic in both cases. **Voice** Assorted harsh guttural croaks and bleats in flight or when alarmed. **Habitat** Frequents a wide variety of wetlands. **Behaviour** Typically feeds in open water, often wading deep and chasing prey energetically; sometimes feeds in small groups. **Status** Vagrant. Single birds recorded in the Khalkh River delta, Dornod province, in June of 1995, 1999 and 2002.

Grey Heron *Ardea cinerea*

90–98 cm

ID The largest Mongolian heron, with a heavier build than very slender Purple Heron and lacking the latter's rich chestnut plumage tones. Could be confused with Common or Demoiselle Cranes, but generally (though not always) flies with neck retracted, not extended as with cranes. Also shows no black on face, throat or upper neck. **Voice** Rasping croak (*khaa-aak*) when taking flight, and other harsh noises in breeding colony. **Habitat** A wide variety of water bodies and wetlands with fish, from large lakes and rivers to desert oases. **Behaviour** Nests colonially (sometimes with cormorants and gulls) on islands or other locations secure from predators; builds nests of sticks in tall dead trees when available but also on low bushes, near ground in reedbeds and on cliffs. **Status** Common breeding visitor or passage migrant, late March to mid-October throughout country, though far less frequently encountered in driest parts of Gobi.

Purple Heron *Ardea purpurea*

78–90 cm

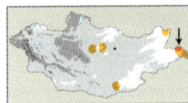

ID In Mongolia, most likely to be confused with much commoner Grey Heron, but strikingly more slender and richly coloured than latter, with boldly striped chestnut head and neck. **Voice** Gives throaty *guwahhh* call in flight. **Habitat** Wetland shallows with reedbeds or other tall vegetation. **Behaviour** Solitary and rather skulking. More active at dawn and dusk. Nests in reedbeds or in bushes and trees when available. **Status** Rare and very local breeding visitor, mid-May to early September; known to nest by Khalkh River, Lake Buir, and possibly Nömrög River, Dornod province. Found at Lake Dashinchilen Tsagaan, Bulgan province, Oyu Tolgoi pond, Ömnögobi province and Lake Ögii, Arkhangai province. **Conservation** Least Concern globally, but considered Near Threatened in Mongolia and listed in the Mongolian Red Book (2013).

breeding

non-breeding

non-breeding

breeding

non-breeding

courtship

Eastern Cattle Egret

Great Egret

breeding

non-breeding

Little Egret

courtship

adult

adult

juvenile

juvenile

Grey Heron

adult

adult

Purple Heron

juvenile

Gyrfalcon *Falco rusticolus* 55–60 cm

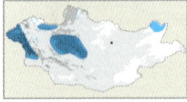

ID A very large and bulky falcon that occurs in three colour forms or 'morphs': white, grey and dark. To date, only the white form has been recorded in Mongolia. Adult white morph is unmistakable. Darker forms could be mistaken for Saker Falcon – especially the large 'Altai Falcon' – which, however, tends to be more brown than grey with a pale crown and cheeks and a thin but distinct 'moustache' stripe, unlike the 'hooded' look of the darkest Gyrs. Juv may be especially confusing since young grey Gyrs can be brownish with paler crown and face and show a weak moustache, while young Sakers can be greyish with a darker crown. **Voice** Loud *kak kak kak* or *kwah kwah kwah* mainly given on breeding grounds. **Habitat** High-altitude barrens and steppes in winter; often associated with rocky terrain. **Behaviour** Feeds on birds and small to medium-sized mammals. **Status** Very rare winter visitor, November to January. The white morph has been reported from Altai and Khangai high mountains and eastern Mongolia, but all records lack documentation.

Saker Falcon *Falco cherrug* 50–58 cm

ID Adult distinguished from Peregrine by larger size, brown (vs. slate-grey) mantle, pale crown and narrow 'moustache' stripe. Tail extends beyond folded wings; flight slower and 'lazier' than Peregrine's. Juv is darker and greyer than adult with more pronounced moustache and thus confusable with young Peregrine, which is browner and more narrowly streaked below. Saker's long tail, dark 'trousers' (leg feathering) and hefty build are best distinctions overall. See also Gyrfalcon. Note: Two subspecies of Saker Falcon occur in Mongolia: *F. c. milvipes*, which occurs in a wide range of habitats throughout the country; and the larger *F. c. altaicus*, restricted to the Altai Mountains of western Mongolia. The latter, sometimes considered a distinct species ('Altai Falcon'), is possibly a relict hybrid population between Saker and Gyrfalcon. **Voice** Gives loud Peregrine-like screaming *keek keek keek*, mainly when breeding. **Habitat** Mountains, forests, wetlands and arid steppe. **Behaviour** Feeds on small mammals and birds. Takes over used nests of other raptors, corvids, etc. in trees, on cliffs, or on ground. **Status** Locally fairly common resident breeder and fairly common passage migrant, mid-April to mid-September; rare winter visitor. Density depends on populations of rodents and passerines. **Conservation** Globally Endangered and Vulnerable in Mongolia. Rapidly declining due to illegal and unsustainable capture for falconry trade, electrocution on power lines, poisoning and other causes. Mongolia remains a regional stronghold. Listed as Near Threatened in the Mongolian Red Book (2013). The national bird of Mongolia since 2012.

Barbary Falcon *Falco pelegrinoides* 43–48 cm

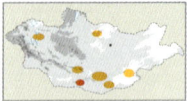

ID Slightly smaller than Peregrine, with more slender build. Paler grey mantle and rufous head markings, when visible, are diagnostic. Juv. also has pale markings on forehead and nape, but some juv. Peregrines are similar. **Voice** Similar to Peregrine. **Habitat** Rocky foothills with some trees or bushes in desert and arid steppe. **Behaviour** Like Peregrine, stoops from height on avian prey. **Status** Rare summer visitor and passage migrant to different parts of Mongolia, e.g. Bayankhongor, Övörkhangai, Uvs and Ömnögobi provinces, mid-April to mid-September. Possible, but unconfirmed breeder in south-west. **Taxonomy** Considered as a subspecies of Peregrine Falcon by BirdLife International.

Peregrine Falcon *Falco peregrinus* 38–48 cm

ID Combination of large size, dark hood with broad 'moustache' stripe, equal wing/tail length and powerful flight, is definitive for adults and juveniles. However, see Barbary Falcon, Gyrfalcon, Saker Falcon and Eurasian Hobby for specific distinctions. **Voice** Shrill, harsh *kreee kreee kreee...* screams given mainly at the nest site. **Habitat** Cliffs and sandy precipices near rivers and lakes with abundant prey. Occurs in arid steppe and Gobi Desert on migration. **Behaviour** Feeds almost exclusively on birds, usually attacking with high-speed stoop. **Status** Very rare and local breeding visitor and uncommon passage migrant, late April to early September across most of Mongolia.

adult
white morph

Gyrfalcon

adult
white morph

adult
grey morph

adult
grey morph

adult

Saker Falcon

adult

Barbary Falcon

juvenile

adult

calidus

peregrinus

peregrinus

adult

adult

adult

Peregrine Falcon

adult

juvenile

Eurasian Hobby *Falco subbuteo* 30–36 cm

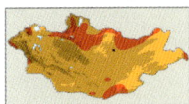

ID Superficially resembles miniature Peregrine, but much more graceful and agile. Bright rufous 'trousers' and undertail of adult are diagnostic. Juv closely resembles juv Amur Falcon but mantle scaling and breast streaking darker (less reddish) and tail only lightly barred above. **Voice** Shrill *klee klee klee* given on breeding grounds. **Habitat** Forest and forested steppe in mountains and river valleys. **Behaviour** Like Amur Falcon, hawks for insects and consumes them on wing; also takes small birds, lizards and rodents. Uses deserted crows' nests. **Status** Fairly common passage migrant and breeding visitor, mid-April to late August throughout Mongolia except for alpine areas and most extreme arid regions of south.

Merlin *Falco columbarius* 25–30 cm

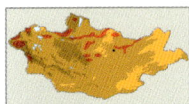

ID Mongolia's smallest falcon, but more compact and 'muscular' than other small falcons, with notably fast and direct flight. In adult male, combination of bluish-grey upperparts and buffy to orange underparts, indistinct moustachial stripe and broad dark band near the end of the tail are diagnostic. Female and juvenile have indistinct 'moustache' and bold tail barring above. **Voice** Shrill and agitated *kikiki* call given on breeding grounds. **Habitat** Nests in abandoned corvid nests in forest and forest steppe of mountains and river valleys, but occurs frequently in steppe and desert on migration. **Behaviour** Hunts mainly small birds in high-speed chases; also takes small rodents. **Status** Uncommon breeding visitor in forested regions but to be expected nearly throughout as passage migrant, late April to late September. Rare visitor in winter.

Red-footed Falcon *Falco vespertinus* 28–34 cm

ID Male distinguished from very similar Amur Falcon by blackish (vs. white) underwing-coverts. Female has orange-buff crown, nape and underparts with little dark streaking, in contrast to female Amur's slate-grey upperparts and boldly streaked underparts. **Voice** Quiet away from roosts and breeding grounds. **Habitat** Open country. **Behaviour** As Amur Falcon. **Status** Vagrant. One record from Khudriin Davaa, Selenge province, undated; also, doubtful records from Hustai Nuruu, Töv province, June 1999; and Khalkh River, Dornod province, June 2000. **Conservation** Near Threatened globally.

Amur Falcon *Falco amurensis* 26–30 cm

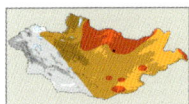

ID Small falcon with long, narrow wings and tail. Male has uniform dark grey plumage with rufous vent and undertail-coverts and striking white underwing-coverts. Female resembles Eurasian Hobby but with less powerful build and 'lighter' manner of flight. Young male is intermediate between female and adult male. Juv distinguished by rufous crown and rufous-buff fringed upperparts. **Voice** A sharp chittering *kikikik* or *kew kew kew* on breeding grounds and at roosts. **Habitat** Forest edges in mountain steppe and near watercourses with scattered trees. **Behaviour** Catches insects on wing (like Eurasian Hobby), but also hovers frequently (like kestrels). Breeds and roosts colonially, using old nests of other raptors and (especially) corvids. **Status** Locally common breeding visitor in forested areas in northern part of country. Very local breeder in eastern Gobi and fairly common passage migrant, early May to late August. [Alt: Eastern Red-footed Falcon]

Lesser Kestrel *Falco naumanni* 29–32 cm

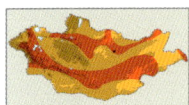

ID Male distinguished from similar Common Kestrel by smaller size and more delicate build. Unspotted mantle, notably pale and sparsely spotted underwings with contrasting dark wing-tips, and less pronounced breast and facial markings. Female and juv very similar to Common Kestrel, though generally less boldly marked. At close range claws pale (vs. black in Common Kestrel). **Voice** Shrill three-note *chee chee chee* flight call given mostly at breeding sites. **Habitat** Nests in open, dry country with hills and cliffs, grassland and semi-desert, but occurs throughout Mongolia on migration. Often uses cavities in artificial structures as well as rock faces. **Behaviour** Colonial nester (unlike Common Kestrel). Hovers, but less regularly than Common Kestrel. Gathers in flocks during migration. **Status** Fairly common breeding visitor and passage migrant, late April to late August. **Conservation** Recently downgraded to Least Concern from Vulnerable.

Common Kestrel *Falco tinnunculus* 32–35 cm

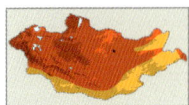

ID Males of the two kestrel species are readily distinguished from all other Mongolian falcons by their reddish-brown mantle and contrasting blue-grey head and tail. Female is also distinctively reddish-brown above, with spotted rather than streaked or barred underparts. For a comparison of kestrels, see Lesser Kestrel (above). **Voice** Noisy when breeding, with piercing screams and chitters; most frequently a high-pitched *kee-kee-kee...* **Habitat** Found in all habitats except for dense taiga forest and alpine zone. **Behaviour** Nests in cavities, including tree-holes, rock crevices, large stick nests of other raptors and corvids, and artificial structures. Frequently hovers in stationary position, dropping to ground to capture small mammals and other small prey. Less gregarious than Lesser Kestrel. **Status** Common and widespread breeding visitor and passage migrant, late April to late September; also a winter visitor. [Alt: Eurasian Kestrel]

Eurasian Hobby

adult

juvenile

juvenile

adult

juvenile

♂

♀

♀

Merlin

♀

♂

♂

♀

♀

♂

Red-footed Falcon

juvenile

♂

♀

juvenile

♂

♀

immature

♂

immature

♀

Amur Falcon

♀

immature

♂

Lesser Kestrel

♂

♀

♂

♀

♀

♂

immature

♀

♂

♂

♀

♂

♀

Common Kestrel

Western Osprey *Pandion haliaetus*   55–58 cm

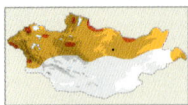

ID Adult's combination of large size, long, narrow wings, short squared tail and mostly white underparts with bold black markings are unlike any other Mongolian raptor. In flight, long, bowed wings and heavy wingbeats give gull-like impression. Juv very similar to adult but has fine bars on tail and lacks black band under wing. **Voice** Short, plaintive whistles. **Habitat** Lakes and rivers with abundant fish. **Behaviour** Commonly soars, circles and hovers over water, and plunges steeply into water to catch fish with talons. Makes large nests of sticks and branches in tall trees near water. **Status** Rare and local breeding visitor and passage migrant, late April to early September. [Alt: Osprey]

European Honey Buzzard *Pernis apivorus*   52–59 cm

ID Very similar to Crested Honey Buzzard, but smaller and lacks crest. Black margins to white throat and mesial stripe of latter are lacking on European Honey Buzzard. Plumage very variable, from light to dark morphs. In flight, paler forms show prominent dark carpal patch at bend of underwing (not present in Crested Honey Buzzard). For comparison with other similar raptors, see Crested Honey Buzzard. **Voice** Drawn-out gull-like whistle. **Habitat** Wooded areas, open country. **Behaviour** Similar to Crested Honey Buzzard. **Status** Vagrant with only two sight records: one in poplar plantation of Tuvshin tourist camp (38 km NW of Dalanzadgad town), Ömnögobi province, 26 May 1998; and one in planted poplars at the Juulchin Gobi tourist resort, Ömnögobi province, on 22 August 1998.

Crested Honey Buzzard *Pernis ptilorhynchus*   M 57 cm, F 60.5 cm

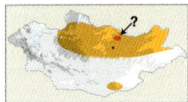

ID A highly variable, broad-winged raptor with dark and light forms and differing degrees of barring on breast, wings and tail. Adult male has dark iris; yellow in adult female. The most distinctive individuals have small crest, boldly barred flight feathers and tail below, and white throat with black borders and mesial stripe. Lacks Eurasian Honey Buzzard's dark carpal patch on underwing. Flight is Accipiter-like: steady flaps interspersed with long glides. Notably long, slender head and extended neck in flight is useful distinction from *Buteo* species. **Voice** Four-note whistle; also drawn-out scream in breeding season. **Habitat** Coniferous and mixed forest. **Behaviour** Builds stick nest in trees. Often feeds on ground, raiding bee and wasp nests for larvae. **Status** Possible rare breeding visitor in Hentii mountain range and rare passage migrant through river valleys of north-east, late April to early September. [Alt: Oriental Honey Buzzard]

Black Kite *Milvus migrans*   56–57 cm

ID Distinguished from other Mongolian raptors by combination of overall dark brown coloration; long, typically bowed wings with pale 'windows' in primaries contrasting with black wing-tips; and long, slightly forked tail and graceful swooping flight. See Booted Eagle and Western Marsh Harrier for specific distinctions. The subspecies that occurs in Mongolia (*M. m. lineatus*) has dark patch through eye and more prominent pale primary patches than nominate form and is sometimes considered a full species, Black-eared Kite. **Voice** Vocal in spring, with quavering, descending *pipipiryoryo* whistle. **Habitat** Occurs in all habitats, including city centres. **Behaviour** Gregarious. Circles and soars at length, often in groups, in lazy cruising flight while scanning below for carrion and human left-overs; also takes some live prey. Nests on virtually all substrates: cliffs, trees and artificial structures. **Status** Very common breeding visitor and passage migrant to all parts of Mongolia, late March to mid-September.

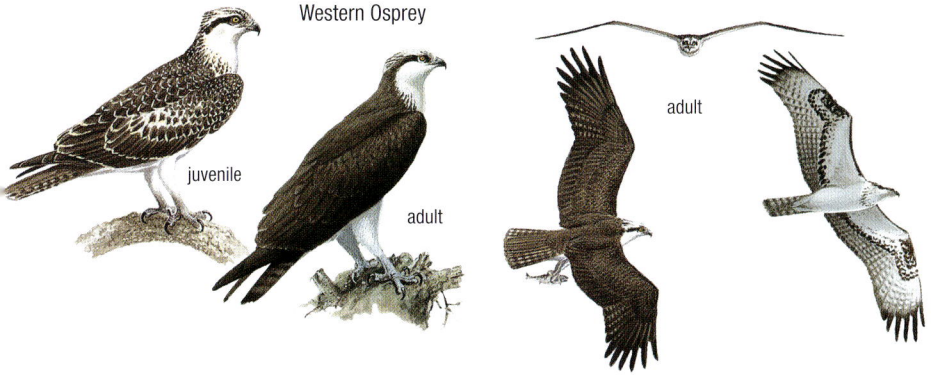

Western Osprey

juvenile

adult

adult

European Honey Buzzard

♂

juvenile
dark morph

♂

♀

juvenile

Crested Honey Buzzard

♀

♂

juvenile

♀

♂

Black Kite

adult

juvenile

adult

juvenile

lineatus

lineatus

Hen Harrier *Circus cyaneus* 44–52 cm

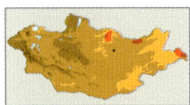

ID As a group, harriers are readily distinguished from other raptors by their long narrow wings, typically raised in a shallow V; long narrow tail; slow, tilting manner of flight; long legs, visible when perched; and small, somewhat owl-like head. All species prefer open country habitats. Adult male Hen Harrier best distinguished from Montagu's by stockier build and absence of black bands across secondaries (above and below) and from Pallid by darker grey mantle and throat, and more extensive black in wing-tips. Adult female very similar to Montagu's and Pallid and best told by build and subtle details of wing markings and facial pattern. Juv Hen Harrier is streaked like female and never has clear rufous breast. Habitat preferences and status also give clues to the identity of these often confusing raptors. For identification of Eastern and Western Marsh Harriers and Pied Harrier, see those species. **Voice** Gives whistling and chattering calls on breeding grounds. **Habitat** Bogs, marshes and shores with emergent vegetation when breeding; also occurs in arid steppe and Gobi Desert on migration. **Behaviour** As with other harriers, courses low over the ground, twisting and turning as it searches for prey. Ground nest in wetland vegetation is made of sticks, reeds and sedges. Takes mainly small birds and mammals. **Status** Very uncommon local breeding visitor and passage migrant throughout Mongolia, late April to early October. Winters in small numbers.

Western Marsh Harrier *Circus aeruginosus* 50–53 cm

ID The two marsh harrier species are larger, with notably broader wings and less buoyant manner of flight, than other harrier species. Adult male Western Marsh in flight has distinctive contrasting back pattern with dark brown mantle, pale grey secondaries and black wing-tips. Adult female is uniform dark brown except for pale buffy crown, throat and 'shoulders'. Juv is like female but pale areas more obscure or (rarely) absent. **Voice** Various whistling or chattering calls on breeding grounds. **Habitat** Mainly reedbeds on margins of lakes and rivers, but occurs in desert steppe and Gobi on migration. **Behaviour** Hunts in typical harrier manner, taking more amphibians and other aquatic prey than upland harriers. Ground nest of sticks, reeds and other wetland plants is sited in dense reedbeds. **Status** Rare breeding visitor in the Uvs and Great Lakes Depression and uncommon passage migrant in northern and western Mongolia, mid-April to early October. Eastern limits of range unclear due to confusion with Eastern Marsh Harrier.

Eastern Marsh Harrier *Circus spilonotus* 48–58 cm

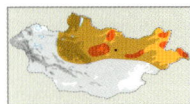

ID Adult male's back pattern in flight similar to that of Western Marsh Harrier, but brown head and mantle of latter replaced by dark grey and black. This creates resemblance to more slender and crisply marked Pied Harrier, which see for more detailed distinctions. Adult female has variegated plumage in contrast to Western Marsh Harrier's rather uniform look, and in flight has distinctive grey flight feathers with bold black barring. Juv resembles juv Western Marsh Harrier. See preceding species for distinctions from upland harriers. **Voice** Similar to Western Marsh Harrier. **Habitat** Mainly reedbeds bordering rivers and lakes. **Behaviour** Similar to Western Marsh Harrier. **Status** Uncommon and local breeding visitor from open wetlands of Hentii mountain range east through Dornod steppe, mid-April to early September. Western extent of breeding range is uncertain due to confusion with Western Marsh Harrier. Also an uncommon passage migrant through breeding range, and possibly further west, as well as through arid steppe and Gobi Desert.

♂

♀

immature

♂

♀

Hen Harrier

♂

♂

juvenile

♀

Western Marsh Harrier

♂

♀

♀

Eastern Marsh Harrier

juvenile

PLATE 20: HARRIERS II

Pallid Harrier *Circus macrourus*

40–50 cm

ID The most lightly built and slim-winged harrier, with notably buoyant flight. Ad male distinguished from Hen and Montagu's by very pale grey upperparts, throat and breast; limited wedge of black in wing-tips; and lack of secondary bars. Female has distinctive underwing pattern with wing bases below darker and with less barring than Hen or Montagu's and a very boldly marked face with pronounced pale collar and dark 'ears' and eye-stripes. Juv has paler wing-tips below, lacking pronounced dark 'fingertips' of Montagu's, and a very bold face pattern as in female. **Voice** Gives *pirrrr* call in display; alarm call a sharp *gik gik gik* on breeding grounds. **Habitat** Open steppe and desert scrub; less associated with wetlands than other harriers. **Behaviour** Nesting and feeding behaviour similar to Hen Harrier. **Status** Rare passage migrant and summer visitor, mainly in western Mongolia, mid-April to early May and late August to mid-September. **Conservation** Considered Near Threatened globally, though Asian populations appear to be more stable than European ones.

Montagu's Harrier *Circus pygargus*

43–47 cm

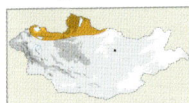

ID Between heavier Hen and lighter Pallid Harriers in overall build. Ad male distinguished from Hen by black secondary bars above and below and from Pallid by darker coloration and more extensive black of wing-tips. Female best told from Hen by slender build, more pointed wings and bold face pattern and from both Hen and Pallid by a distinctive broad pale band across the secondaries below. Juv from Pallid by more uniform wing coloration below, and less bolder face pattern. **Voice** Various whistling and chittering calls on breeding grounds. **Habitat** Open country, mostly close to river and lake valleys with reedbeds. **Behaviour** Feeding and nesting behaviour as in Hen Harrier. **Status** Rare passage migrant and summer visitor to Northern Uvs and Darkhad Depression, mid- to late April and late August to early September. Nesting unconfirmed.

Pied Harrier *Circus melanoleucos*

41–46.5 cm

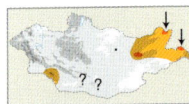

ID Bold, clean-cut, black-and-white plumage of ad male confusable only with male Eastern Marsh Harrier, which has a heavier build, more restricted black on wing-tips below and a pattern of black speckling rather than solid black on head and mantle – though in some individuals head and upper back are black. Adult female is also 'cleaner' and lighter than Eastern Marsh Harrier with a uniform dark brown back and pale grey wings with bold black striping and pale grey, lightly barred tail. Juv is similar to adult female Eastern Marsh Harrier and juv Montagu's Harrier, but differs by dark brown upperparts; broader, clear white rump patch; and lack of pale markings on forewing and tail. **Voice** Lapwing-like call on territory, also chattering *chak chak chak* alarm call. **Habitat** Open country, mostly in river and lake valleys with reedbeds. **Behaviour** as Hen Harrier. **Status** Rare and very local breeding visitor in eastern Mongolia and rare passage migrant in central and eastern Mongolia, mid-April to mid-September.

immature ♂

Pallid Harrier

♂   ♂

♀

♀

♀

juvenile

Pallid juvenile

Montagu's juvenile

Montagu's Harrier

♂ immature

juvenile

♂

♂

♀   ♀

♀

Pied Harrier

♂

♂

♀

♂ juvenile

♀

Northern Goshawk *Accipiter gentilis* M 49–56 cm; F 58–64 cm

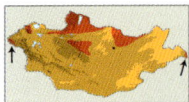

ID Best distinguished from all other Mongolian accipiters by its large size (nearly twice as large as the sparrowhawks) and bulky build, comparable to that of buzzards. Distinguished from all other regularly occurring raptors in flight by combination of long, rounded, broadly barred tail and relatively short wings. From vagrant Mountain Hawk-Eagle by finely grey-barred underparts (adult male and female) and dark tear-drop breast streaking (juv). Juv plumage pattern is similar to juv Saker Falcon, but has creamy-buff underparts, broad and heavily barred wings, distinctly barred tail, and no moustache. **Voice** Quiet away from breeding grounds, where it gives a deep chattering *kya kya kya* or *keee keee keee*. **Habitat** Forest, but recorded in steppe and desert during migration. **Behaviour** Hunts in forest, using cover to ambush prey. Performs soaring display in early spring. Hunts medium-sized birds (e.g. grouse) and mammals, especially squirrels. Defends nest aggressively against intruders, including people. **Status** Fairly common, though typically secretive breeding visitor in forested mountains and taiga zone and uncommon passage migrant throughout, mid-April to mid-October. A few adult birds overwinter.

Eurasian Sparrowhawk *Accipiter nisus* M 29–34 cm; F 35–41 cm

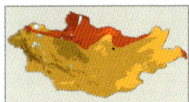

ID Mongolia's three small accipiters are easily distinguished from all other raptors in flight by short, rounded wings and long tail, resulting in characteristic flight pattern of rapid flapping, alternating with longer glides. This species is notably larger than relatively rare Japanese Sparrowhawk. Male has rufous cheeks and breast barring. Female and juv are grey-barred below (not rufous-barred or streaked). Vagrant Shikra is strikingly pale in all plumages. Northern Goshawk is much larger, bulkier and more ponderous in flight. **Voice** Gives shrill chattering call at nest, otherwise quiet. **Habitat** Forests on mountainsides, often near rivers and lakes. Migrates across entire country, including steppe and desert. **Behaviour** Hunts small songbirds (e.g. tits and thrushes), rarely small mammals inside woodlands. Soars on fine days and during migration. **Status** Fairly common breeding visitor and common passage migrant, late March and late September. Rare but irregular in winter.

Shikra *Accipiter badius* 30–36 cm

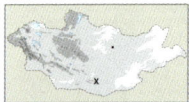

ID Much paler in all plumages than either Eurasian Sparrowhawk or rarer Japanese Sparrowhawk, and far smaller than Northern Goshawk. Both male and female are grey above with fine rufous barring below. Juv is similar to Eurasian Sparrowhawk but has dark streaking (vs. barring) below. Both ad and juv have prominent mesial stripe. In flight, adults have only faint barring on underwings (accentuating dark wing-tips) and tail. **Voice** Quiet away from breeding grounds. **Habitat** Open country with trees. **Behaviour** Similar to Eurasian Sparrowhawk. **Status** Vagrant. A single sight record near Juulchin Tourist Camp, Ömnögobi province, 29 May 1986.

Japanese Sparrowhawk *Accipiter gularis* M 27 cm; F 30 cm

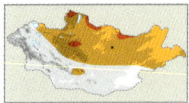

ID Notably smaller than more common Eurasian Sparrowhawk and vagrant Shikra and much darker above and below than latter. Adult male lacks rufous cheeks of Eurasian Sparrowhawk. Adult female has dark rufous, not grey, barring below. Juv has combination of streaked breast and barred belly, unlike other small accipiters. **Voice** Chattering *kyik-kee-kee-kee* given at nest and also on migration. **Habitat** Nests mainly in mountain forest, often near rivers or lakes. **Behaviour** Like other small accipiters, feeds mainly on small woodland birds ambushed in flight. **Status** Very rare and local breeding visitor in mountain ranges of north and east Mongolia and passage migrant across most of country, mid-April to mid-September.

♀

juvenile

♂

♀

Northern Goshawk

juvenile

♂

♂

♀

juvenile

♂

♀

♂

♀

Eurasian
Sparrowhawk

♂

♀

Shikra

juvenile

juvenile

Japanese Sparrowhawk

♀

juvenile

juvenile

♀

♂

♂

Rough-legged Buzzard *Buteo lagopus*   49–51 cm

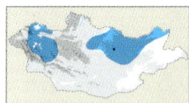

ID Adult distinguished from all other Mongolian buzzards by white tail with contrasting black terminal bands both above and below (adult female and juv have single broad band; adult male 3–4 bands). Dark morph Common and Long-legged Buzzards have a dark terminal tail-band, but inner tail is never bright white. Proportionately longer wings than other buzzards (except Long-legged) create a distinctive flight profile. Occurs in winter only. **Voice** Typically silent on wintering grounds. **Habitat** Nests in Arctic; winters in open steppe and river valleys, where it finds thin snow cover and an abundance of small rodents, especially Brandt's Voles. **Behaviour** Hovers much more frequently while hunting than other buzzards. **Status** Uncommon to fairly common winter visitor near Selenge, Orkhon and Tuul Rivers, Great Lakes Depression and similar open regions. Early October to early March, numbers fluctuating with population cycles of rodent prey. [Alt: Rough-legged Hawk]

Upland Buzzard *Buteo hemilasius*   57–62 cm

ID Adult shares most plumage characteristics and variations (dark, pale and rufous) with other buzzard species, which see for key distinctions. Most readily distinguished by its notably large size and bulky appearance; also by feathered tarsi ('trousers'), uniform white and slightly rufous tail with a few dark bars near tip; and distinctive dark belly patch. Juv is very similar to adult, but has pale iris. In flight, large size plus longer tail and wings suggest small eagle. Compare also Rough-legged Buzzard (winter). **Voice** Typical *Buteo* mewing calls. **Habitat** Occurs in most habitats throughout Mongolia, though prefers steppe and desert (absent from taiga forest and alpine zone above 3,000 m). **Behaviour** Soars and hovers with wings held in a distinct V. Feeds mainly on small mammals. **Status** Common resident breeder, occupying nest sites as early as late March. Fairly common in winter, with most adult birds wintering near active Brandt's Vole colonies. In all seasons, populations fluctuate with rodent densities. Some juveniles apparently migrate south in winter, departing from late September to mid-October. [Alt: Mongolian Buzzard]

Long-legged Buzzard *Buteo rufinus*   61 cm

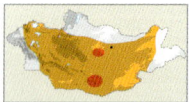

ID As with other buzzard species, plumage highly variable. In the most distinctive plumages, presents very clean appearance with creamy-buff or reddish-brown head and breast and pale, nearly unmarked tail (never barred on belly like Common and Eastern Buzzards). Dark morphs resemble Rough-legged Buzzard but tail markings less distinct and Long-legged does not occur in winter. Larger and longer-necked than Common and Eastern Buzzards, with longer wings and tail, and soars with wings in deeper V, yet notably smaller and less bulky than Upland Buzzard. **Voice** Gives loud mewing call. **Habitat** Open, rocky valleys, steppe and (mainly) desert. **Behaviour** Like other *Buteo* species, soars in search of small mammal (mostly rodent) prey. **Status** Rare and local breeding visitor and uncommon passage migrant, mid-April to early September. Only one confirmed nest at Khongoryn Els, Ömnögobi province, July 2009 and 2015, with young seen in same area the following August.

Common Buzzard *Buteo buteo*   48–56 cm

ID The smallest buzzard in Mongolia. Very similar to Eastern Buzzard (see below) and identification criteria not fully established (other than by range). **Voice** As Eastern Buzzard. **Habitat** As Eastern Buzzard. **Behaviour** As Eastern Buzzard. **Status** May nest in conifer forest in western Mongolia.

Eastern Buzzard *Buteo japonicus*   51–59 cm

ID Small and compact buzzard, with relatively short wings and short, broad tail; soars with wings held flat (vs. V-pattern of other buzzards). Plumages are highly variable and *B. buteo* and *B. japonicus* may not be safely distinguishable in field, except perhaps by range. The paler forms of both are separable from Long-legged Buzzard by barred sides and (usually) belly; from Upland Buzzard by small size, slighter build and habitat preference; and from Rough-legged Buzzard by lack of black-and-white tail pattern and seasonality. Dark forms are best distinguished by size, build and behaviour. Juv resembles adult, but has paler iris, more barring on tail, and more streaking below. **Voice** Calls frequently in all seasons, usually in flight, a loud mewing *piioou*. **Habitat** Forest, nesting mainly in coniferous trees, though also on cliffs in forested valleys. Uses all habitats in Mongolia on migration. **Behaviour** Feeds on small mammals (mainly rodents) and other prey caught on ground; also takes carrion. Soars and glides at length and sometimes hovers. **Status** Common breeding visitor in forested regions of Mongolia and a common passage migrant countrywide, mid-April to mid-September. [Alt: Japanese Buzzard]

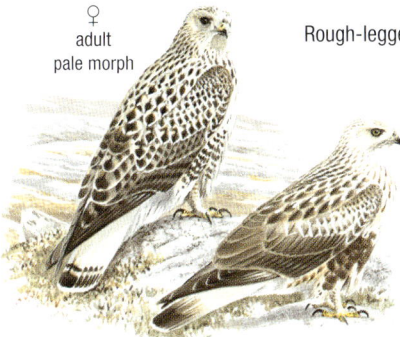

♀
adult
pale morph

Rough-legged Buzzard

juvenile

♀
adult

juvenile

adult

juvenile

adult
dark morph

Upland Buzzard

juvenile

adult

adult
dark morph

juvenile

Long-legged
Buzzard

adult
pale morph

adult
rufous morph

adult

adult

buteo

adult
dark
morph

adult

adult

japonicus

adult
rufous morph

Common Buzzard

Eastern Buzzard

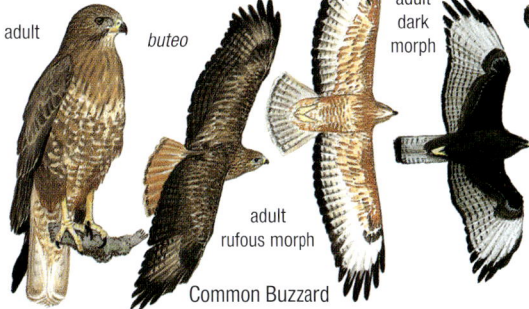

Grey-faced Buzzard *Butastur indicus*   41–48 cm

ID Looks like mixture of buzzard, accipiter and harrier due to longish tail and *Buteo*-like body. This 'hybrid' look is a clue to identity. However, the adult's combination of white throat with black margins and mesial stripe, grey face and white supercilium are diagnostic. See also Crested Honey Buzzard. Juv is brown above with streaked breast, and dark face mask with bold supercilium. In flight, resembles large Eurasian Sparrowhawk. **Voice** A whistled *pik-wheee* or *kin-meee* heard all year but especially prior to nesting. **Habitat** Open country with wooded areas. **Behaviour** Hunts small mammals, snakes, lizards and insects. **Status** Vagrant. Included here on basis of single unconfirmed sight record at the Khonin Nuga research station in west Hentii Mountain, Selenge province, June 1999.

Mountain Hawk-Eagle *Nisaetus nipalensis*   66–84 cm

ID Unlikely to be mistaken for any other Mongolian raptor due to powerful build; distinctively shaped broad wings with 'secondary bulge' in flight; generally pale underparts; and habitat preference. Adult is heavily barred and streaked below. Juv has clear whitish breast, belly and underwing-coverts. **Voice** Usually quiet, but may give assorted piping whistles or screaming calls. **Habitat** Mountain forest and forested river valleys. **Behaviour** Preys mainly on medium-sized mammals and birds. **Status** Vagrant. A single sight record in the Tuul River valley of Songino, near Ulaanbaatar, 14 June 1995 (lacks documentation). **Taxonomy** Formerly placed in genus *Spizaetus*. [Alt: Hodgson's Hawk-Eagle]

Booted Eagle *Hieraaetus pennatus*   42–51 cm

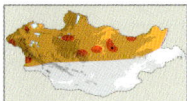

ID A small, buzzard-sized eagle with long 'fingers' and eagle-like flight. The pale morph is very distinctive from below, with black flight feathers contrasting with otherwise whitish underparts. All plumages show broad, pale wing patches above. Pale wing patches, long square-ended tail (which can appear slightly forked), overall rich brown coloration and manner of flight while hunting can give dark morph bird a strong resemblance to much more common Black Kite; however, Booted Eagle always shows a pair of white spots ('landing lights') on leading edge of wing on either side of neck. **Voice** Generally quiet, but has buzzard-like mew and emphatic shrill *kik-kik-kik* chatter. **Habitat** Forest, including taiga, as well as mixed forest of mountains and river valleys. **Behaviour** Hunts small mammals, reptiles and birds. Performs spectacular headlong stoops to the ground. **Status** Uncommon and local breeding visitor to the northern, forested half of the country and uncommon passage migrant, mid-April to early September. **Taxonomy** Sometimes placed in genus *Aquila*.

Bonelli's Eagle *Aquila fasciata*   55–67 cm

ID This very rare, medium-sized eagle readily distinguished from other Mongolian raptors by the following features: adults of both sexes have white breast with fine tear-drop streaks, contrasting with dark flight feathers, and unique black bar extending from body to bend of underwing. From above, the notably broad wings, whitish mantle patch and broad black terminal band on tail are diagnostic. Juv can be largely white or warm brown below; possible confusion with Booted Eagle, but distinguished by larger size, broader wings, rounded tail and strongly protruding head. Imm resembles juv but has black carpal patch and streaks on underparts. **Voice** Silent away from breeding grounds, where it gives a soft fluty call. **Habitat** Open mountain country with cliffs and trees. **Behaviour** A versatile hunter, taking medium-sized birds and mammals by ambush, low-level quartering or stooping from height. Often nests on cliffs. **Status** Vagrant. Two sight records: one at Khan Bogd, Ömnögobi province, 3 September 1987, and one by Bulgan River, Khovd province, 17 September 1996. **Taxonomy** Formerly placed in genus *Hieraaetus*.

adult

juvenile

adult

Grey-faced Buzzard

juvenile

adult

adult

Mountain Hawk-Eagle

adult

adult
dark morph

adult
dark morph

adult

Booted Eagle

adult
pale morph

adult

juvenile

juvenile

adult

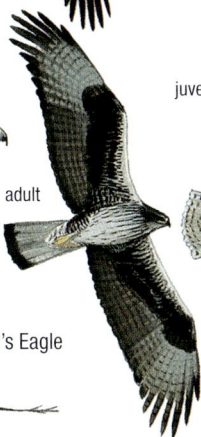

Bonelli's Eagle

Steppe Eagle *Aquila nipalensis*   62–74 cm

ID By far the most common and widespread eagle in Mongolia. Most likely to be confused with rare Greater Spotted Eagle, which is slightly smaller and more compact with a smaller bill. At close range when perched, Steppe Eagle's prominent yellow gape, extending to rear edge of eye or just beyond, is diagnostic. In flight, Steppe shows longer wings and 'fingers' and longer tail than Greater Spotted; has distinct barring on flight and tail feathers below; and usually shows marked contrast below between paler brown wing-coverts and darker flight feathers. Most Steppe Eagles also show white 'windows' at base of primaries above. Juv has pale edges to wing-coverts, a pale crescent at the base of the tail, a pale-tipped tail and prominent 'baggy trousers' (feathered tarsi). See also Golden and Imperial Eagles for relevant comparisons. **Voice** On breeding grounds gives hoarse barking *kow kow kow* and various croaks and whistles. **Habitat** Steppe and mountain ranges, with marked preference for open country. **Behaviour** Feeds on wide variety of items, including prey as large as marmots, Corsac Fox and domestic goats, as well as small birds, rodents and even insects; also feeds on carrion and follows herds of gazelles during calving and migrations. **Status** Common breeding visitor and passage migrant throughout Mongolia, mid-April to early October; very rare and local in winter, where large rodent populations are found. **Conservation** Globally Endangered.

Greater Spotted Eagle *Clanga clanga*   59–69 cm

ID Smallest, darkest, and most compact and uniformly coloured large eagle in Mongolia. Most likely to be confused with much more common Steppe Eagle, which has different build, markings and habitat preferences, as detailed under that species. Juv Greater Spotted has distinctive white spots on mantle. See also Golden and Imperial Eagles. **Voice** Abrupt yelping *kyu kyu kyu* call on territory and in winter. **Habitat** Breeds in forested areas near rivers, lakes and wetlands, nesting in large trees. **Behaviour** Hunts mammals as well as waterbirds, snakes and frogs. **Status** Very rare and local breeding visitor and passage migrant in northern Mongolia, mid-April to early October. **Conservation** Classified as globally Vulnerable and Endangered in Mongolia. Listed as Threatened in the Mongolian Red Book (2013). **Taxonomy** Formerly placed in genus *Aquila*. [Alt: Spotted Eagle]

Eastern Imperial Eagle *Aquila heliaca*   72–83 cm

ID This rare eagle most resembles the much more common Golden Eagle, with relatively large size, long wings and tail and flat wings in flight distinguishing it from other *Aquila* species. Adult is shorter-tailed, with greyer (vs. rich brown) plumage overall than Golden. Nape patch paler (more silver than gold) than Golden, and white shoulder patches are diagnostic, though these can be small and obscure. Juv is boldly patterned with dark flight and tail feathers strongly contrasting with pale sand-coloured body and wing-coverts; this combination most resembles juv Steppe Eagle, but note Eastern Imperial's different coloration, streaked breast and absence of barring in flight feathers. Imm distinctively streaked against otherwise pale plumage. **Voice** Loud quacking barks *yowp yowp yowp* given on territory and elsewhere. **Habitat** Forest in mountains and river valleys. **Behaviour** Preys on small to medium-sized birds and mammals on ground. **Status** Very rare breeding visitor in Hentii and Hövsgöl mountain ranges (a single documented nest site in 1990) and rare passage migrant elsewhere. Satellite tracking has shown that birds from eastern Russian population pass through Mongolia en route to wintering grounds in China. Occurs late April to late October and possibly during winter (single sight record from Hentii province, February 1987). **Conservation** Considered Vulnerable globally and in Mongolia due to small and declining populations. Listed as Threatened in the Mongolian Red Book (2013).

Steppe Eagle

adult

juvenile

juvenile

adult

juvenile

Greater Spotted Eagle

adult

adult

juvenile

adult

juvenile

Eastern Imperial Eagle

adult

adult

adult

juvenile

juvenile

Golden Eagle *Aquila chrysaetos* — 80–93 cm

ID This is the largest of the *Aquila* eagles in Mongolia and is readily distinguished by its build and manner of flight as well as plumage characteristics. Despite its large size it is relatively long-winged and long-tailed and flies with its wings held in a shallow V, all of which give it a more buoyant, graceful appearance in the air than its congeners. The white tail with broad black terminal band is diagnostic in juvs and subadults, and young birds also show conspicuous white 'windows' at base of outer flight feathers. The tail in adults is grey above with prominent black bars and broad terminal band, unlike that of other eagles, except for rare Eastern Imperial, which see for further distinctions. **Voice** Various yapping and whistling calls. **Habitat** Forested mountains, steppe and desert with cliffs and high rocky slopes. **Behaviour** Takes prey up to size of foxes and young deer and gazelles, and also takes medium-sized and large birds, and occasionally reptiles during breeding season. Feeds on carrion in non-breeding season. Nests in trees, on cliffs and among boulders close to ground, building several large stick nests within breeding territory and defending them year round. Captured and used for traditional Kazakh falconry in western Mongolia. **Status** Uncommon to fairly common resident breeder.

Pallas's Fish Eagle *Haliaeetus leucoryphus* — 76–84 cm

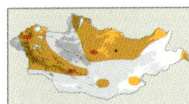

ID Adult readily distinguished from other large eagles by whitish head and upper breast contrasting with dark brown plumage above and below; dark bill; and (especially in flight) largely white tail with broad black terminal band. Juv is mostly dark brown showing bold white flashes in the underwing and pale wing-coverts in flight. **Voice** Noisy on breeding grounds, with harsh, throaty, barking *kyow kyow kyow*; also gull-like calls. **Habitat** Lakes and rivers with abundant fish. **Behaviour** Ponderous in flight and spends much time resting. Feeds on fish and less frequently waterbirds, in both cases preferring dead, sick or young individuals to active, healthy ones. **Status** Very rare and local breeding visitor and passage migrant, late April to mid-September. No recent breeding record in Mongolia. **Conservation** Considered globally Vulnerable. Listed as Threatened in the Mongolian Red Book (2013). Mongolia is a stronghold for this rare eagle, though there is evidence of recent decline. [Alt: Pallas's Sea-eagle]

White-tailed Eagle *Haliaeetus albicilla* — 70–90 cm

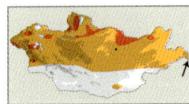

ID In all plumages, bulky build, very short wedge-shaped tail (contrasting with unusually broad, rectangular wings in flight) and heavy bill distinguish this huge, somewhat vulture-like eagle from all other raptors. Yellow bill and white, unmarked tail of adult are diagnostic. Juv largely blackish-brown with restricted pale areas increasing as it matures over five years; pale centres to blackish tail feathers below are diagnostic. Compare Pallas's Fish Eagle. **Voice** Gives various gentle barks and yelps on the breeding grounds, also more emphatic alarm calls. **Habitat** Close to large rivers and lakes in both forested regions and open country, though absent from desert. **Behaviour** Spends long periods perched on rocks, frozen lakes or in trees, scanning for prey. Feeds mainly on fish and waterfowl. **Status** Uncommon and local (though widespread) breeding visitor and uncommon passage migrant, late April to early October. A few birds winter where waterfowl are attracted to open water. Largely absent from very arid south. **Conservation** Recently down-graded from Threatened to Least Concern due to recent increases in much of its range. Listed as Near Threatened in the Mongolian Red Book (2013). [Alt: White-tailed Sea-eagle]

Golden Eagle

adult

adult

adult

adult

juvenile

Pallas's Fish Eagle

adult

adult

juvenile

immature

juvenile

adult

immature

juvenile

adult

adult

juvenile

adult

White-tailed Eagle

Short-toed Snake Eagle *Circaetus gallicus* 62–67 cm

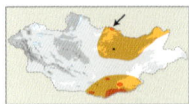

ID A large, long-winged and notably large-headed raptor. Distinguished from other Mongolian raptors of similar size by white underparts boldly striped and barred with dark brown, and contrasting dark head. The palest Eastern Buzzards can be similarly marked but are much smaller. Juv is even paler below (as well as above) and may have white throat. **Voice** Shrill, clear, whistling two-note mew. **Habitat** Open, dry areas with trees. **Behaviour** Hunts snakes, which it pounces upon from perch or (often) hovering flight. Often chooses prominent perching spots. **Status** Rare and very local breeding visitor with recent nesting records from Selenge River valley and in Ömnögobi province. Uncommon passage migrant through river valleys of the north-east (Hentii range), and presumably elsewhere, late April to early May and late August to early September. **Conservation** Listed as Threatened in the Mongolian Red Book (2013). Widely distributed in Europe and Central Asia, but several populations are declining. [Alt: Short-toed Eagle]

Bearded Vulture *Gypaetus barbatus* 108–115 cm

ID Enormous but relatively slender compared to other large vultures and readily distinguished in flight in all plumages by long, pointed wings, notably long, wedge-shaped tail and pointed, protruding head and neck. Tawny head and underparts contrasting with silvery-black upperparts are diagnostic in perched adult. Blackish juvenile best told from other vultures (when perched) by long tail and small head. Juv of vagrant Egyptian Vulture shares some characteristics but is much smaller. **Voice** Usually silent, though may give shrill whistles at nest. **Habitat** Rocky valleys in high mountains where wild and/or domestic mammals present in numbers. **Behaviour** Much more buoyant and agile in flight than other vultures. Not gregarious. Feeds mainly on carrion and bones, dropping the latter from height to break them and access the marrow. Makes huge stick nest in crevices and caves. **Status** Rare and local (though often conspicuous) resident breeder in Hentii, Mongol-Altai and Gobi-Altai mountain ranges. **Conservation** Considered Near Threatened globally and Vulnerable in Mongolia due to limited distribution and recent declines. Listed as Threatened in the Mongolian Red Book (2013). [Alt: Lammergeier]

Egyptian Vulture *Neophron percnopterus* 55–65 cm

ID Overall white plumage, with contrasting black flight feathers of adult, may be confused with similar pattern of Himalayan Vulture. The latter is twice as large, however, with much bulkier build, much heavier bill and black tail. In flight, juv Egyptian might be confused with juv Bearded Vulture, which also has slender wings, a wedge-shaped tail and a narrow head, but again the latter dwarfs Egyptian Vulture and has longer wings and tail. **Voice** Usually silent, but rarely high-pitched mewing near the nesting site. **Habitat** Open country near high mountains. **Behaviour** Subordinate to larger vultures when gathered at carrion, its main diet. Also eats eggs. **Status** Vagrant with only three sight records: one between Gurvansaikhan mountain range and Borzongiin Gobi, Ömnögobi province, in August 1980; one at Mönkh Khairkhan Mountain, Khovd province, 24 May 1996; and one in Dungenee valley of Gurvansaikhan Mountains, Ömnögobi province, 24 June 1998. **Conservation** Considered Endangered globally due to recent steep declines in much of its range due to pesticide poisoning and other causes.

adult

juvenile

gliding

soaring

adult
pale individual

juvenile
pale individual

Short-toed Snake Eagle

adult

juvenile

adult

juvenile

Bearded Vulture

adult

adult

adult

immature

immature

adult

immature

juvenile

Egyptian Vulture

Cinereous Vulture *Aegypius monachus*

100–115 cm

ID An enormous black, soaring bird with broad rectangular wings, readily distinguished in all plumages from similar-sized Griffon Vulture and Himalayan Vulture by is uniform dark coloration. Soars with wings held flat and tips often lowered. Juv has dark head, neck and ruff. **Voice** Hisses when feeding, gives throaty cackle when taking off. **Habitat** Rocky cliffs and valleys of high mountains and adjacent steppes where medium to large wild and domestic mammals are present in numbers. **Behaviour** Soars and glides over wide range, sometimes at great height, in search of carrion. Dominates other species when feeding. Makes massive stick nests on rocky substrates and trees when available. **Status** Common resident breeder, often in numbers, and uncommon passage migrant, March to September. Adult birds winter in Mongolia; young tagged birds have been recorded wintering in South Korea, Nepal and China. **Conservation** Considered Near Threatened globally due to steep declines in parts of it range. [Alt: Black Vulture]

Griffon Vulture *Gyps fulvus*

95–105 cm

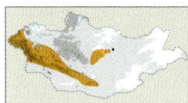

ID Adult similar to Himalayan Vulture, which see for detailed distinctions of all plumages. **Voice** Gives quiet grunts and hisses at roost and when feeding. **Habitat** Rocky habitats in high mountains where wild and/or domestic mammals present in numbers. A few recent records from steppe and forest steppe. **Behaviour** Highly gregarious. Feeds on carrion and spends much time on the wing searching for it. **Status** Very rare non-breeding summer visitor and passage migrant, late April to early September. [Alt: Eurasian Griffon]

Himalayan Vulture *Gyps himalayensis*

125–150 cm

ID Very similar to Griffon Vulture but larger and 'chestier'. Adult best distinguished by whitish-buff (not brownish-grey) mantle and underparts with less pronounced breast streaking. Juv is blackish with prominent grey streaks, quite unlike much paler juv Griffon. At close range, perched Himalayan has pinkish (not grey) legs and feet and yellowish (not blackish) cere. **Voice** Similar to Griffon Vulture. **Habitat** Cliffs and rocky valleys in high mountains with adjacent steppe and numerous wild and/or domestic mammals. **Behaviour** Similar to Griffon Vulture. **Status** Common non-breeding summer visitor and passage migrant, late April to early September. **Conservation** Recently upgraded to Near Threatened globally due to recent declines in its main (Himalayan) range caused by pesticide poisoning. [Alt: Himalayan Griffon]

Cinereous Vulture

juvenile

adult

adult

juvenile

adult

adult

juvenile

juvenile

adult

Griffon Vulture

adult

juvenile

juvenile

adult

Himalayan Vulture

adult

juvenile

Brown-cheeked Rail *Rallus indicus*   23–26 cm

ID Relatively long, slender red bill distinguishes this species from other members of its family in Mongolia. The paler, less colourful juv can also be identified by bill length. **Voice** A distinctive piglet-like squeal is the most familiar of its many calls; when breeding, males make a rhythmic series of *kip* notes. **Habitat** Reedbeds, sedge meadows or other dense wetland vegetation along lake and river margins. **Behaviour** Very shy and skulking, more often heard than seen. In cold weather, more inclined to venture along shores with sparse vegetation and reedbed margins. When flushed, flies a short distance with legs dangling before dropping back into vegetation. Feeds on aquatic invertebrates and plants. **Status** Uncommon and local breeding visitor and rare passage migrant, mid-April to early September; possibly more common than records indicate due to secretive nature. **Taxonomy** Formerly treated as a subspecies of Water Rail *Rallus aquaticus*. [Alt: Eastern Water Rail]

Spotted Crake *Porzana porzana*   19–22.5 cm

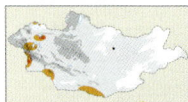

ID Dense white spotting on face, neck and breast, and prominent buffy undertail with few (or no) markings separate all plumages from other short-billed rails in Mongolia. Adults have relatively deep bill, orange at base with yellow tip. Juv similar but markings less distinct. **Voice** Breeding male gives a distinctive, far-carrying, repeated *huitt*, often compared to crack of whip or 'wolf whistle'. **Habitat** Dense sedgy marshes (but not reedbeds) on margins of lakes and rivers. **Behaviour** Usually remains hidden within its dense habitat while nesting, and more often heard than seen. Easier to observe on migration. **Status** Very rare and local breeding visitor and passage migrant, late April to early September; known from only a few localities in western Mongolia and Gobi.

Little Crake *Porzana parva*   17–19 cm

ID Adult male very similar to Baillon's Crake, but duller brown above, lacking significant black-and-white barring on flanks, and has red bill base. Long wings extend well beyond tail. Distinguished from other short-billed crakes by slate-blue face and underparts. Female and juv are whitish-buff and relatively unmarked on breast and belly – very unlike other female and juv crakes in Mongolia. **Voice** Breeding males give a loud, abrupt barking *kua* in an accelerating series. **Habitat** Favours reedbeds. **Behaviour** Very reclusive and hard to see, especially at breeding sites. **Status** Very rare and local breeding visitor and passage migrant; recorded only at Lake Khar-Us in western Mongolia, but may nest elsewhere in Great Lakes Depression, late April to early September.

Baillon's Crake *Porzana pusilla*   16–18 cm

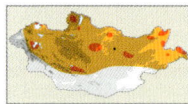

ID Very similar to much rarer Little Crake, but ad male and female are warmer, more rufous-brown above, have prominent black-and-white barring on flanks and lack a red bill base. Distinguished from other Mongolian crakes by deep blue-grey underparts. Juv is buffy-brown below, but much more heavily barred on underparts than juv Little Crake. **Voice** Territorial male gives a rather weak, dry frog-like trill or rattle *kokoko* or *tou tou tou*; a sharp *tack* call is given in alarm. **Habitat** Marshes, sedgy bogs and meadows, and sometimes reeds along lake and river edges. **Behaviour** Like other rails, reclusive in its dense habitat and best located by voice. **Status** Uncommon and local, but widespread breeding visitor and passage migrant, mid-April to early September. Recorded at a number of sites from Great Lakes Depression in western Mongolia to boreal Hövsgöl region and arid areas in Valley of the Lakes, all the way to far eastern Dornod province.

Swinhoe's Rail *Coturnicops exquisitus*   13 cm

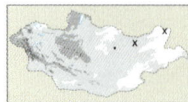

ID Could be confused with imm Little or Baillon's Crakes, but in addition to tiny size may be recognised by distinctive bold striping running from nape to tail and on flanks, and (in adults) all-yellow bill. White secondaries, conspicuous in flight, are diagnostic. **Voice** Silent away from breeding grounds. **Habitat** Wet grasslands and marshes in river and lake valleys. **Behaviour** Rodent-like in its preference for concealing itself in dense grasses and other vegetation; will risk being stepped upon rather than flying. **Status** Vagrant. Known from only two records: one at Lake Tari or Tooroi, Dornod province (undated); and first-year bird in valley of Khurkh River, Hentii province, 8 September 2002. **Conservation** Considered globally Vulnerable. [Alt: Swinhoe's Crake]

adult

Brown-cheeked Rail

juvenile

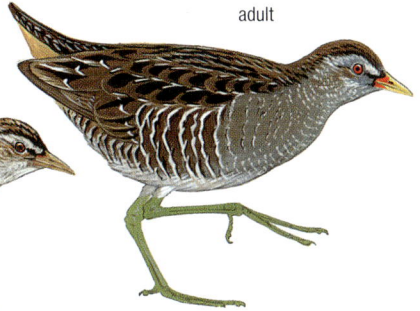

adult

Spotted Crake

♀

juvenile

♂

Little Crake

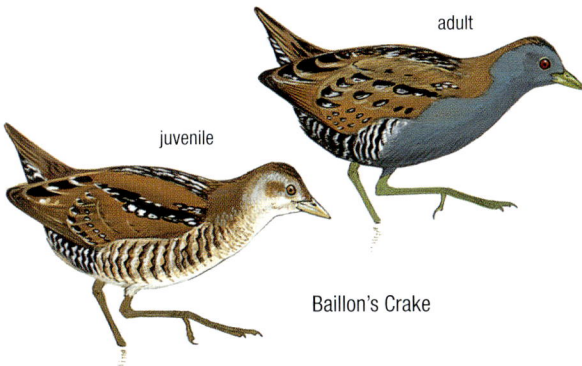

juvenile

adult

Baillon's Crake

adult

Swinhoe's Rail

PLATE 29: CORNCRAKE & GALLINULES

Corncrake *Crex crex*   22–25 cm

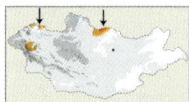

ID Stocky crake with stout pinkish to reddish bill and legs. Superficially resembles juv or female Little Crake or juv Baillon's Crake, but distinguished by rufous-and-white barred flanks and chestnut wings; also (typically) by habitat preference. **Voice** Song is a mechanical, grating *crex crex*, repeated endlessly at night. **Habitat** Wet and dry grassland; less typically, reedbeds. **Behaviour** Remains hidden in long, dense grass, so difficult to see. Occurs in drier habitats than other crakes. **Status** Very rare breeding visitor and passage migrant. Recorded as breeding only in valley of Zelter River (Selenge province), Northern Uvs Depression and once (presumably as a migrant) in plantation in Khovd city, Khovd province, 31 May 2006.

White-breasted Waterhen *Amaurornis phoenicurus*   28–33 cm

ID Unmistakable. A large, moorhen-like rail, slate-grey above contrasting strongly with white face and underparts; undertail-coverts cinnamon; bill bright yellowish-green with red spot at base. **Voice** Noisy, with assorted croaking, chuckling and gulping calls. **Habitat** Dense vegetation near water. **Behaviour** Not shy. Restless, flicking cocked tail. **Status** Vagrant. Only three records: a dead bird in concrete enclosure at Tsaidam well in Darkhan district, Hentii province, 8 May 2004; remains of one near Three Camel Lodge, Ömnögobi province, 2 June 2004; and a single bird at Khan Bogd, Ömnögobi province, 1 May 2013.

Common Moorhen *Gallinula chloropus*   27–35 cm

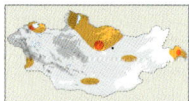

ID Adults distinguished from all other waterfowl by dark slaty-blue plumage, yellow-tipped red bill and forehead 'shield', white flank stripe and white 'flags' on cocked tail. For similarities with Eurasian Coot, see that species. **Voice** Vociferous, with abrupt shrill liquid *pruuuk* and other harsher calls. **Habitat** Marshes and reed-edged ponds and lakes. **Behaviour** Swims and also forages on land. Shier and less gregarious than Eurasian Coot. Flicks tail constantly. **Status** Uncommon and local but widespread breeding visitor and passage migrant, late April to early September.

Eurasian Coot *Fulica atra*   36–42 cm

ID Adult distinguished from all other waterfowl by striking white bill and forehead shield. Juv resembles juv Common Moorhen, but has white throat, neck and breast and dusky undertail, while latter has the opposite colour combination. **Voice** Noisy, with a sharp *kik kik* and *krek* among its many calls. **Habitat** Reed-edged wetlands, lakes and ponds; occurs in oases in arid steppe and desert on migration. **Behaviour** Duck-like, swimming and diving for aquatic plants, but also foraging along shores; often occurs in flocks. **Status** Fairly common breeding visitor and passage migrant throughout most of Mongolia, late April to early September. [Alt: Common Coot]

Corncrake

adult

adult

juvenile

White-breasted Waterhen

Common Moorhen

adult

juvenile

Eurasian Coot

adult

juvenile

Red-crowned Crane *Grus japonensis*

138–152 cm

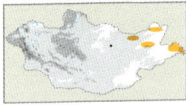

ID Combination of white body with black neck, face and tail 'bustle', red crown patch and yellow bill are diagnostic among Mongolian cranes. Imm similar, but with dark rusty-brown (not black) on head and neck and no red crown patch. **Voice** Loud trumpeting and bugling calls given in breeding season; quiet *kruip* call on take-off and landing; imm has whistling contact call. **Habitat** River and lake valleys with reedbeds. **Behaviour** Gregarious. Feeds mainly on small vertebrates and large insects captured on the ground. **Status** Very rare summer visitor and passage migrant in eastern Mongolia (especially Dornod province), late May to early September. **Conservation** Considered Endangered globally and listed in the Threatened Birds of Asia (2001). [Alt: Japanese Crane].

Siberian Crane *Leucogeranus leucogeranus*

120–140 cm

ID No other crane species is wholly white (standing) with bright red bill, face and legs. Imm has brownish bill and is strongly marked with cinnamon-brown on head, neck, mantle and wings. In flight, both adult and imm show contrasting black wing-tips. Superficially similar Oriental Stork has much longer black bill and more extensive black in wings (visible when standing); smaller Great Egret has yellow bill and black legs. **Voice** High-pitched *koonk koonk* call in flight; imm has whistling contact call. **Habitat** Lakes and riverbanks with reedbeds and other wetlands. **Behaviour** Forages in steppe and wetlands for large insects, small vertebrates and plant matter; flocks with other cranes in winter and on migration. **Status** Very rare passage migrant and summer visitor to northern parts of country, early May to early September. **Conservation** Critically Endangered globally and in Mongolia due to its limited range and declining population. Listed in the Threatened Birds of Asia (2001) and Mongolian Red Book (2013). **Taxonomy** Formerly placed in genus *Grus*.

Common Crane *Grus grus*

110–120 cm

ID Most likely to be confused with smaller Demoiselle Crane (Plate 31), but black of neck does not extend to breast; white on face is more extensive; white neck stripe is longer; and it has prominent tail 'bustle' (thin and pointed in Demoiselle). Imm has brown markings on upperparts, with buff or grey head and neck. **Voice** Gives loud trumpeting note in flight and noisy bugling duets on breeding grounds. **Habitat** Nests in marshy edges and reedbeds in lake and river valleys with scattered shrubs and small trees (especially willows); also on agricultural land. Occurs in desert steppe on migration. **Behaviour** Food preferences include aquatic tubers, molluscs and insects, frogs and toads, but – like all cranes – essentially omnivorous. Family groups remain together on migration and in winter. **Status** Fairly common breeding visitor and passage migrant (in flocks of up to 500) across northern and central Mongolia, though absent from much of arid south; early April to mid-September. **Conservation** Considered stable across its broad Eurasian breeding range but at risk in Mongolia due to widespread habitat degradation. Listed as Near Threatened in the Mongolian Red Book (2013).

adult

juvenile

Red-crowned Crane

adult

adult

juvenile

Siberian Crane

adult

juvenile

Common Crane

White-naped Crane *Antigone vipio*
125–153 cm

ID Combination of large stature, white pattern on head and neck, large red face patch, pale yellow bill and reddish legs, and whitish tail 'bustle' is unique among Mongolian cranes. Imm similar, but with rusty-brown on head and neck, and wings more grey-brown. In flight, all flight feathers are blackish, strongly contrasting with pale grey upperwing- and underwing-coverts. **Voice** Low-pitched duetting between pairs begins on wintering grounds. Loud bugling calls in flight. **Habitat** Nests near lakes and in river valleys and wetlands with tall vegetation and reedbeds; visits wheat fields and dry steppe on migration. **Behaviour** Like other cranes, feeds on small vertebrates, large invertebrates, plant seeds and tubers. Forms large flocks away from breeding grounds, mixing with other cranes. **Status** Rather rare and local breeding visitor, mainly in north-east, but also in scattered localities in Great Lakes Depression and central Mongolia, mid-April to mid-September. **Conservation** Considered Vulnerable globally and in Mongolia due to limited breeding and wintering range and a number of threats to its population. Listed as Threatened in the Threatened Birds of Asia (2001) and Mongolian Red Book (2013). **Taxonomy** Formerly placed in genus *Grus*.

Hooded Crane *Grus monacha*
91–100 cm

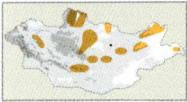

ID A small, dark crane. Adult has white head and neck, small red and black forecrown patch, and blackish legs – unlike any other Mongolian crane. Also, the only crane with no contrast on upperwing in flight. Imm lacks black and red on forehead and has rusty off-white head and neck and somewhat browner plumage than adult. **Voice** Gives loud *kereeek* or *kerreeer* in flight; imm has whistling contact call. **Habitat** Wetlands and grasslands, including farmland and lake beds. **Behaviour** Like other cranes, feeds on terrestrial and aquatic invertebrates, vertebrates and plant matter. Recorded in large flocks (to 1,000 birds) with other crane species in wheat fields during autumn migration. **Status** Regular, but very uncommon and local, summer visitor and passage migrant, mainly in north-east, but also in scattered wetland habitats in central Mongolia, late April to early September. Possibly breeds in isolated wetland habitat in Darkhad Depression (Hövsgöl province) and Hentii. **Conservation** Considered Vulnerable globally and in Mongolia due to restricted breeding and wintering range and multiple threats to population. Listed as Threatened in the Threatened Birds of Asia (2001) and Mongolian Red Book (2013).

Demoiselle Crane *Grus virgo*
90–100 cm

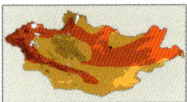

ID The smallest Mongolian crane. Adult is most likely to be confused with Common Crane, but white tuft behind eye (vs. facial and neck stripe); grey crown (without red patch); black neck feathers extending below breast in a loose tuft; and pointed tertials (vs. 'bustle') are diagnostic. Imm is like adult, but head and neck dark grey, eye and chest tufts reduced or absent, with grey-brown cast to upperparts. In flight, Demoiselle is best separated at a distance from Common Crane by black-centred breast. **Voice** Shrill trumpeting on breeding grounds and in winter; also a rolling trill. Imm has whistling contact call. **Habitat** Breeds in steppe and desert-edge habitats; visits wetlands and farmland on migration. **Behaviour** Forages for large insects, small vertebrates and plant matter. Flight faster, more agile than larger cranes. **Status** Common breeding visitor and passage migrant throughout Mongolia (scarcest in driest southern extremities), mid-April to early December. **Taxonomy** Formerly placed in genus *Anthropoides*.

adult

White-naped Crane

courtship display

juvenile

adult

adult

late 1st-winter

adult

Hooded Crane

juvenile

adult

juvenile

adult

Demoiselle Crane

PLATE 32: BUSTARDS, STONE-CURLEW & PAINTED-SNIPE

Great Bustard *Otis tarda*
M 90–105 cm, F 75–85 cm

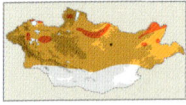

ID Large size, stocky build, and rufous coloration distinguish this species from all bustards but Macqueen's, which is smaller, less 'chunky' and sandy-yellow in colour with a diagnostic black neck stripe. Female is smaller and paler than breeding male. In flight, resembles a long-necked eagle. **Voice** Male has low-pitched *uhmb* call when displaying; alarm call is a short barking note; otherwise silent. **Habitat** Nests in dry valleys of lakes and rivers, mountain steppe and fields; occurs in desert steppe on migration, and winters in wheat fields depending on snow coverage. **Behaviour** Males perform spectacular displays in communal leks, expanding their chest and fanning wings and tail. Feeds on large insects and small mammals, birds, lizards, frogs, seeds and other plant matter. **Status** Uncommon and local breeding visitor and passage migrant throughout the country, though generally absent from the arid far south, where Macqueen's Bustard normally ranges; mid-April to early October and rare winter visitor. Winters in wheat fields in northern Mongolia in small numbers. **Conservation** Considered Vulnerable globally and in Mongolia. Listed as Threatened in the Mongolian Red Book (2013).

Macqueen's Bustard *Chlamydotis macqueenii*
55–65 cm

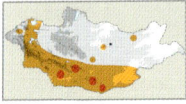

ID In all plumages distinguished from the superficially similar Great Bustard by black stripe running from nape to the breast, though this is most prominent in breeding males. Macqueen's is also smaller and more lightly built than Great Bustard with paler, yellowish (vs. reddish) coloration. Sharp black-and-white wing-tip pattern is diagnostic in flight. At close range, pale eyes produce staring expression. **Voice** Practically silent. **Habitat** Semi-desert with scattered scrub and sandy grassland. **Behaviour** Male performs flamboyant courtship display, withdrawing its head into expanded neck and breast plumage. Omnivorous, foraging on foot for large insects, small vertebrates and plant material. **Status** Rare and local breeding visitor and uncommon passage migrant from Great Lakes Depression and south Khovd province in west across the arid south to south-eastern Dornogobi province, late April to late September. **Conservation** Considered Vulnerable globally and in Mongolia. Listed as Threatened in the Mongolian Red Book (2013). **Taxonomy** Once widely considered a race of Houbara Bustard *C. undulata*. [Alt: Asian Houbara]

Eurasian Stone-curlew *Burhinus oedicnemus*
40–44 cm

ID A stocky, long-legged, large-headed wader of dry habitats with yellow legs, a short thick bill, a bold wing pattern at rest as well as in flight, and unusually large yellow eyes; this combination is unlike any other Mongolian species. **Voice** A curlew-like *coor-lee*. **Habitat** Desert and arid steppe. **Behaviour** Most active dusk to dawn. **Status** Vagrant. One near Lake Khar-Us, Khovd province, June 1993. [Alt: Eurasian Thick-knee]

Greater Painted-snipe *Rostratula benghalensis*
23–28 cm

ID Adult plumage colour and pattern highly distinctive with female brighter than male. Differs from true (*Gallinago*) snipes by bold black-and-white 'braces', boldly barred back, white patches around eyes, and greenish-yellow bill and legs; feet project well beyond tail in flight. **Voice** Females give hooting calls on breeding grounds. Both sexes rarely give a sharp *tooick* or *twick*. **Habitat** Well-vegetated wetlands. **Behaviour** Shy and secretive; active mainly at dawn and dusk. **Status** Vagrant. One on shore of Lake Tsagaan, Bayankhongor province, June 1922 and one near Lake Uvs, Uvs province, June 2013.

Great Bustard

♀

♂

♂
immature

adult

Macqueen's Bustard

adult

Eurasian Stone-curlew

♀
display

♂

♂

♀

Greater Painted-snipe

Grey Plover *Pluvialis squatarola* 27–31 cm

ID Black breast and belly of breeding plumage unique except for Pacific Golden Plover, which is spangled golden-brown above rather than grey and white. Imm closely resembles young Pacific Golden Plover, but distinguished by larger size, heavier bill and – in flight – black axillaries ('armpits') and white rump. **Voice** A sweet three-note *plee-whoo-ee*. **Habitat** Shallow wetlands, riverbanks and lake shores. **Behaviour** In Mongolia, typically found with flocks of Pacific Golden Plovers. Forages with short dashes and pauses typical of plovers. **Status** Uncommon passage migrant in appropriate habitat throughout Mongolia, late April to early May and late August to early September. [Alt: Black-bellied Plover]

Pacific Golden Plover *Pluvialis fulva* 21–25 cm

ID In breeding plumage, resembles Grey Plover, but has gold-spangled (vs. grey-and-white) upperparts. Winter and imm birds distinguished by notably smaller head and bill and (most easily) in flight by uniformly dark rump and tail, and absence of black axillaries ('armpits'). **Voice** Call is a plaintive *chu-vit*, recalling Spotted Redshank flight call. **Habitat** Favours short grass steppe, including burned areas, but also lake shores and wetland edges. **Behaviour** May form large migrating flocks. **Status** Fairly common passage migrant throughout Mongolia in appropriate habitat, late April to early May and mid-August to early September; generally more common in east of country.

Common Ringed Plover *Charadrius hiaticula* 17–19 cm

ID Distinguished from similar Little Ringed Plover by bright orange legs and bill and lack of strong yellow eye-ring (breeding). Imm has heavier bill and strong white (not buff) supercilium. All plumages show clear white wing-stripe, lacking in Little Ringed Plover. **Voice** A fluty whistling *tyoo-li*; alarm call is a sharp *skreeet*. **Habitat** Lake shores, riverbanks and wetland edges where migrant waders gather. **Behaviour** Typical plover run-and-stop feeding behaviour. **Status** Rare passage migrant, late April to early May and late August to early September, with most records coming from Buir Lake-Khalkh River-Khyangan Region in far east, Lake Dashinchilen Nuur, Bulgan province in central Mongolia and Lake Khar-Us and Tes River in the west.

Little Ringed Plover *Charadrius dubius* 14–17 cm

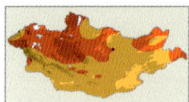

ID Breeding adults best told from migrant Common Ringed Plover by strong yellow eye-ring, thin dark bill and pinkish (not orange) legs. Thin bill and buff (not white) supercilium are best indicators for imm. **Voice** A drawn-out soft *peeeoo*; also gives a trilling *pipipipi* in display on breeding grounds. **Habitat** Dry, sandy or rocky shores near water. **Behaviour** Males perform bat-like flight display, and adults distract predators from nests by feigning injury. **Status** Common breeding visitor and passage migrant throughout most of Mongolia; absent as a breeder in extremely arid south. Nesting birds arrive in late April and early May and depart in late August and early September.

Kentish Plover *Charadrius alexandrinus* 15–17.5 cm

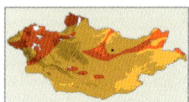

ID Male is smaller and paler with more limited black head and neck markings than other 'ringed' plovers, with a distinctive rufous crown and nape. Female and imm best told by dark bill and leg colour and by habitat. **Voice** Hard *trrrrt* trill given in flight, extended to a rattling *gregregregre* in display flight. **Habitat** Favours open shores, flats and lake beds with scant vegetation, especially alkaline bodies, as nest sites. **Behaviour** Similar to the other 'ringed' plovers. **Status** Common breeding visitor and passage migrant throughout Mongolia; does not nest in the driest areas of the Gobi. Breeders arrive in late April and early May and depart in late August and early September.

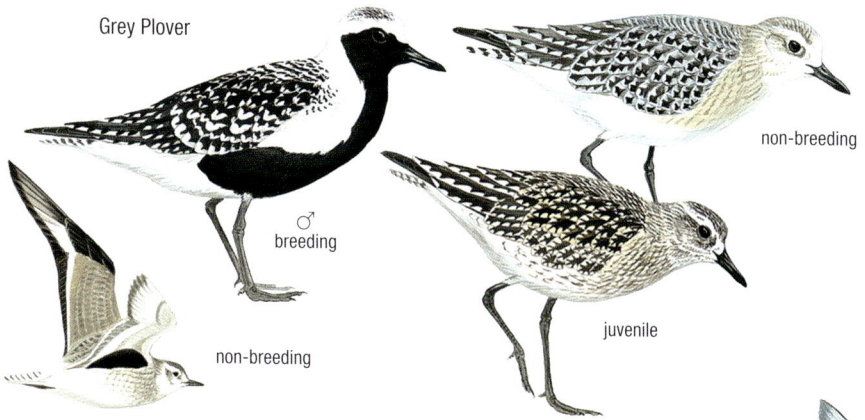

Grey Plover

♂ breeding

non-breeding

non-breeding

juvenile

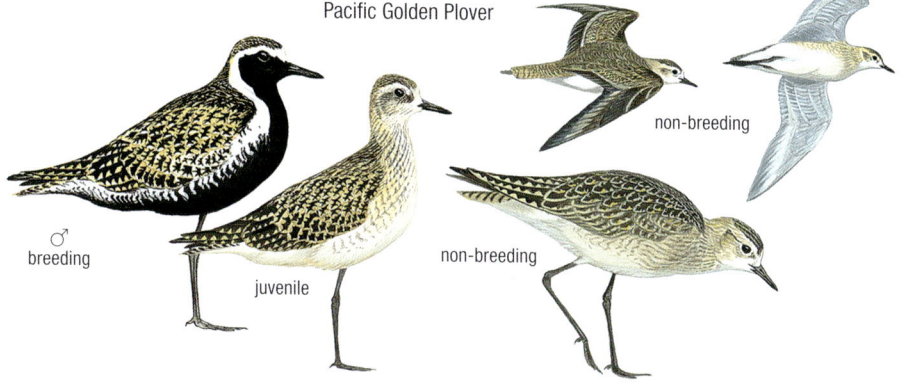

Pacific Golden Plover

♂ breeding

juvenile

non-breeding

non-breeding

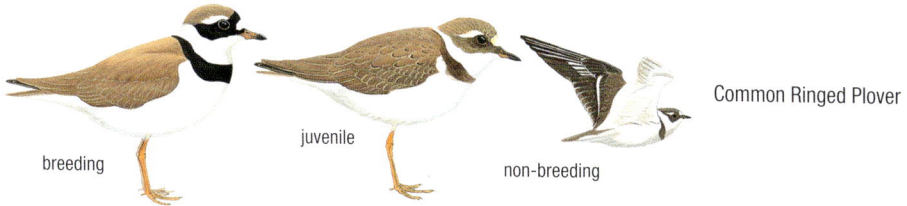

breeding

juvenile

non-breeding

Common Ringed Plover

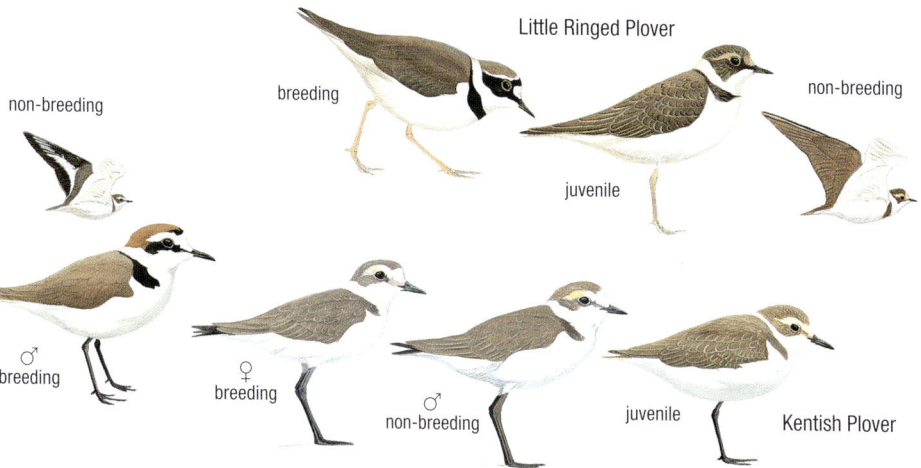

Little Ringed Plover

non-breeding

breeding

non-breeding

juvenile

♂ breeding

♀ breeding

♂ non-breeding

juvenile

Kentish Plover

Greater Sand Plover *Charadrius leschenaultii*   22–25 cm

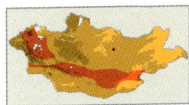

ID Very similar in all plumages to Lesser Sand Plover, from which it is best distinguished by larger size and chunkier build; longer, heavier bill; longer, paler legs (greenish-buff vs. dark grey). Plumage features variable in both species and often unreliable for identification. **Voice** A trilling *trrrrrt* or *prrrrit* in flight, a little lower-pitched than Lesser Sand Plover's call but often hard to distinguish without experience. **Habitat** Breeds on dry, sparsely vegetated steppe. On migration, occurs on lake shores and other habitats where waders congregate. **Behaviour** Reputedly averages more steps (3–10) between pauses while foraging than Lesser (1–4). **Status** Uncommon breeding visitor and fairly common passage migrant throughout Mongolia, late April to late August.

Lesser Sand Plover *Charadrius mongolus*   18–21 cm

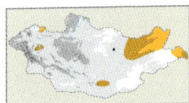

ID In all plumages, difficult to distinguish from Greater Sand Plover, which see for most reliable differences. Distinguished from Kentish Plover by buff rather than white hind-neck, different proportions and larger size. **Voice** Flight call is a soft trill. **Habitat** Lake shores, riverbanks and other wetland habitats where migrant shorebirds gather. **Behaviour** Similar to other sand plovers; however see under Greater Sand Plover. **Status** Uncommon passage migrant, late April to early May and late August, most commonly in far east. **Taxonomy** Sometimes considered to comprise two species: Lesser Sand Plover *C. atrifrons* (with ssp. *C. a. schaeferi*, a passage migrant in southern Mongolia), and Mongolian Plover *C. mongolus* (with ssp. *C. m. mongolus*, a passage migrant in the north). The precise status of these forms in Mongolia is unclear. [Alt: Mongolian Sand Plover]

Oriental Plover *Charadrius veredus*   22–25 cm

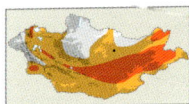

ID Readily distinguished from other sand plovers by white head and bright rufous breast with black lower border (breeding male); rufous-buff breast contrasting with white throat and broad supercilium (breeding female and non-breeding adults). Imm like Lesser Sand Plover but with markedly scaly mantle. In all plumages, an elegantly proportioned species with notably long neck, legs (yellowish-pinkish) and wings. Absence of wing-stripe obvious in flight. Similar Caspian Plover *C. asiaticus* (not recorded in Mongolia, but might occur in the west) has white axillaries and white underwing-coverts. **Voice** Gives loud *chip chip chip* call in flight; other calls more whistling. **Habitat** Dry grasslands. **Behaviour** Males perform a dramatic zooming courtship flight, suggesting a 'lunatic nightjar'. **Status** Fairly common but local breeding visitor to southern and eastern Mongolia and uncommon passage migrant throughout in favourable habitat, late April to mid-August. [Alt: Oriental Dotterel, Eastern Sand Plover]

Eurasian Dotterel *Charadrius morinellus*   20–23 cm

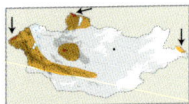

ID Combination of grey neck and upperparts, bright rufous breast, black belly patch, bold white breast-band, and clean white supercilium meeting in a V at the nape make breeding adults unmistakable. Non-breeding adults and imms are greyish-tan with scaly mantle, but retain the conspicuous breast-band and supercilium. In flight, shows no wing-stripe in any plumage. **Voice** Gives a low trilling *prrooot* in flight, sharp double whistle in alarm. **Habitat** Breeds on mountain tundra. On migration, usually found on dry grasslands, occasionally wetland edges. **Behaviour** Eggs incubated mainly by males. **Status** Very rare and local breeding visitor restricted to highest peaks of the Mongol-Altai, Khangai and Hövsgöl ranges, depending on weather conditions. Presumably present late April to late August. Also rare passage migrant with small groups sometimes found with other waders at lower elevations.

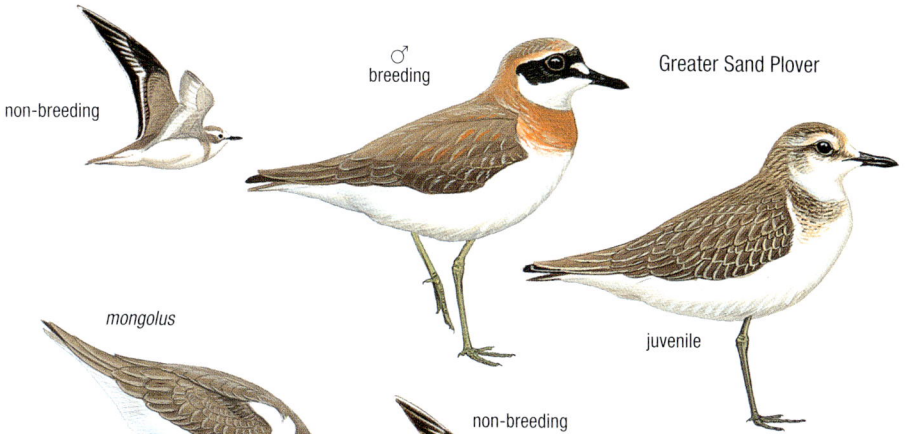

non-breeding

♂ breeding

Greater Sand Plover

juvenile

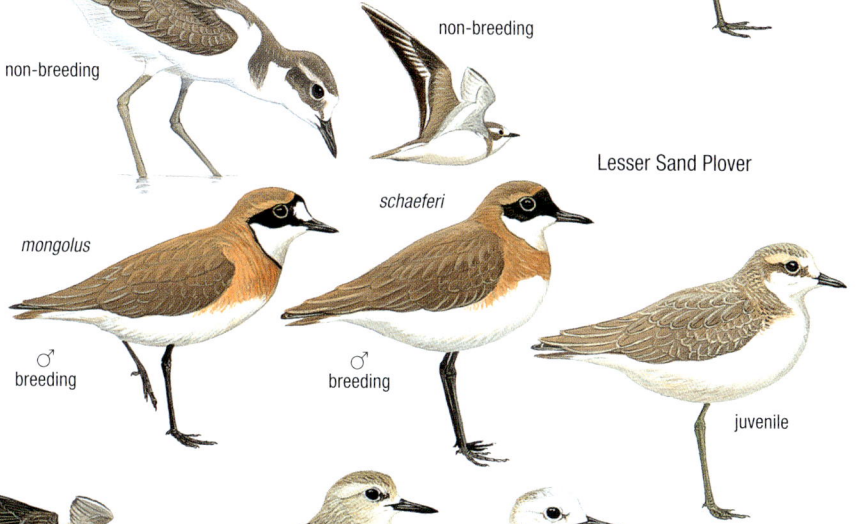

mongolus

non-breeding

non-breeding

Lesser Sand Plover

mongolus

schaeferi

♂ breeding

♂ breeding

juvenile

♀

juvenile

Oriental Plover

juvenile

♂ breeding

juvenile

non-breeding

Eurasian Dotterel

juvenile

♀ breeding

Sociable Lapwing *Vanellus gregarius*   27–30 cm

ID In Mongolia, this large, rare plover is likely to be confused only with similarly rare Grey-headed Lapwing, which lacks this species' black cap, contrasting white supercilium and black belly (breeding adult) and has yellow (vs. black) legs. Winter adult and imm Sociable retain traces of diagnostic head pattern. In flight, from below black belly of adult Sociable distinct from white belly of Grey-headed. **Voice** Call is a harsh *krech*, though generally silent away from breeding grounds. **Habitat** Breeds in dry grassland; migrants may frequent wetland edges or fields with flocks of other *Vanellus* plovers on migration. **Behaviour** Nests semi-colonially. **Status** Vagrant to Mongolia, late April to early May and mid to late August. Mongolian records include: female at Lake Ögii, Arkhangai province, 28 July 1998; male and 2 females at Lake Ögii, Arkhangai province, 16 August 1998; one near Nömrög River, Dornod province, in June 2002; an adult at Lake Tsogiin Tsagaan, Dornod province, in June 1995; and at Khongoryn Els, Ömnögobi province. **Conservation** Critically Endangered with rapid recent population decline projected to continue, possibly due to increased hunting pressure during migration. [Alt: Sociable Plover]

Northern Lapwing *Vanellus vanellus*   28–31 cm

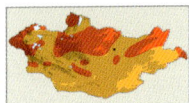

ID Dark, iridescent green/black/purple upperparts, black breast, long wispy crest, broadly rounded wings in flight and orange legs make this species unmistakable. Non-breeding adult has less boldly patterned head and buffy neck. Juv has very short crest and is more buff-fringed above. **Voice** In territorial display, gives a striking toy-trumpet whooping and mewing notes; also a frequent *pee-wit* call; highly vocal in defence of nest and young. **Habitat** Wet grassland, marshes, agricultural fields and wetland edges; also lakes and rivers in arid habitats during migration. **Behaviour** Territorial and breeding displays include aerobatic rolling and tumbling. Highly gregarious after breeding season. **Status** Common breeding visitor in appropriate habitat throughout the country; absent only from driest parts of Gobi and highest elevations. Common to abundant passage migrant, late April to late August. **Conservation** Considered Near Threatened globally.

Grey-headed Lapwing *Vanellus cinereus*   34–37 cm

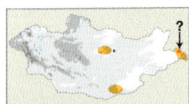

ID In Mongolia confusable only with vagrant Sociable Lapwing, but distinguishable in all plumages by grey head and neck; bright yellow legs and bill with dark tip; and black breast-band and white (vs. black) belly. In flight, shows distinctive black primaries and white underparts, and white tail with black subterminal band. **Voice** Gives loud and sharp *kik kik* calls in flight, and mournful *chi-yit chi-yit* alarm call. **Habitat** Wet grasslands and marshes. **Behaviour** Like other lapwings, tends to be conspicuous and vocal, especially when nesting. Typically forages in shallow water. **Status** Rare passage migrant, late April to early May and late August to early September in eastern, central and southern Mongolia. May nest in valleys of Khalkh and Nömrög Rivers, in Dornod province, though breeding is not confirmed. [Alt: Grey-headed Plover]

Ruddy Turnstone *Arenaria interpres*   21–26 cm

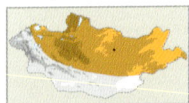

ID Bold black-and-white head and breast pattern and bright chestnut mantle of breeding adult diagnostic. Juv and non-breeding adult duller, but readily recognised by squat form, orange legs, short, plover-like bill and striking upperwing pattern in flight. **Voice** When flushed, gives harsh rattling *tuk tuk-tuk*. **Habitat** Favours rocky lake shores, riverbanks and sandy beaches; less frequently in other open wetland habitats and short grass. **Behaviour** Often feeds by flipping over stones and other objects to obtain invertebrates sheltering underneath. **Status** Fairly common passage migrant in small numbers across northern Mongolia east to Dornod province, late April to early May and late August.

Black-winged Stilt *Himantopus himantopus*   35–40 cm

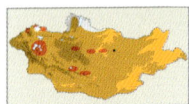

ID Unmistakable, with extremely long, pink legs, black-and-white plumage and fine straight bill. Juv is browner above. Compare Pied Avocet. **Voice** A loud shrill *kik kik kik* or *kyip kyip kyip* given on breeding grounds. **Habitat** Typically, shallow waters at edges of marshes, pools and lakes. **Behaviour** Often feeds in small parties; may wade into deep water, snapping up flies and other prey from surface. Drives away intruders near nest with fierce aerial mobbing. **Status** Uncommon and local breeding visitor and uncommon passage migrant along lake shores and rivers across Mongolia, though absent from driest parts of Gobi and highest mountains, late April to early September.

Pied Avocet *Recurvirostra avosetta*   42–45 cm

ID Superficially resembles Black-winged Stilt, but distinguished by black cap and nape, long, upturned bill and blue-grey legs; otherwise unmistakable. Juv is browner and more mottled above. **Voice** A loud fluty *klee klee klee* or *kweet kweet kweet*. **Habitat** Open, sandy shores and mudflats as well as marsh edges and banks of rivers and lakes; more tolerant of desert habitats than Black-winged Stilt, especially during migration. **Behaviour** Often wades deeply and swims. Sweeps bill rapidly from side to side in water when feeding. Colonial nester, very noisy and often aggressive to other birds. **Status** Uncommon and local breeding visitor and rare passage migrant throughout Mongolia except highest mountain regions, late April to early September.

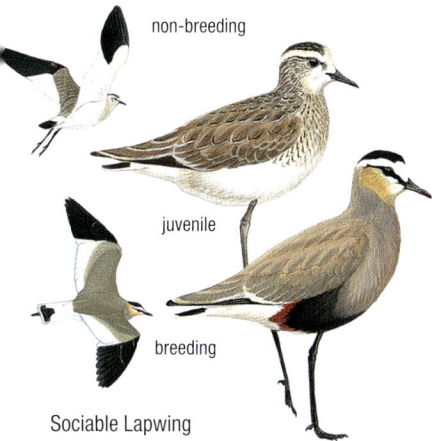

non-breeding

juvenile

breeding

Sociable Lapwing

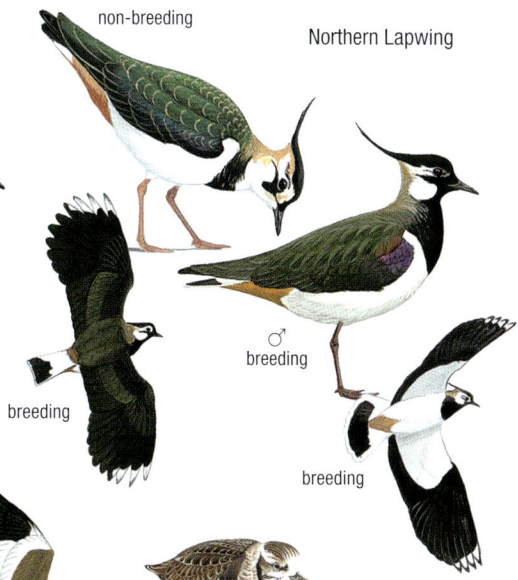

non-breeding

Northern Lapwing

breeding

♂ breeding

breeding

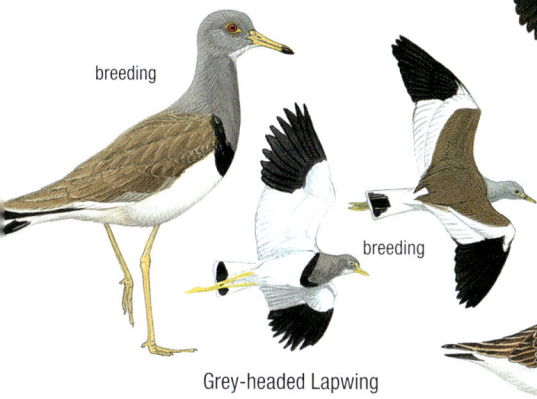

breeding

breeding

Grey-headed Lapwing

non-breeding

juvenile

♂ breeding

Ruddy Turnstone

♂

♀

♂

Black-winged Stilt

adult

Pied Avocet

PLATE 36: SNIPES & WOODCOCK

Jack Snipe *Lymnocryptes minimus* — 18–20 cm

ID A notably small, short-billed snipe, both secretive and rare in Mongolia. In addition to size and bill length, the best features are: double supercillium; absence of pale central crown-stripe; and very prominent yellowish mantle stripes. **Voice** Quiet, but may give low soft *guh* or *gah* when flushed. **Habitat** Wetlands. **Behaviour** Shy and skulking. Moves jerkily, like clockwork toy. Freezes rather than flushes, but if flushed, flies off silently in straight line. **Status** Rare passage migrant; recorded only at Lake Orog, Bayankhongor province, 10 October 1971 and Lake Buir, Dornod province (date unknown).

Common Snipe *Gallinago gallinago* — 23–28 cm

ID Very similar to Pin-tailed and Swinhoe's Snipes, but can be distinguished by white trailing edge to secondaries and white patch and stripe on underwings when flushed; comparatively narrow supercillium vs. broad dark eye-stripe at base of bill; moderate foot projection in flight; longer bill (than Swinhoe's); and erratic, zig-zag flight when flushed. **Voice** Call when flushed is a harsh *skaiak* or *jaaak*. On breeding grounds gives incessant *chippa-chippa-chippa* notes from ground or raised perch. **Habitat** Breeds in damp, marshy or boggy areas in all regions. On migration, uses a wide range of open wetland types. **Behaviour** Discreet feeder, often keeping close to cover. Probes damp ground for invertebrate prey. In display, circles in deeply undulating flight, producing bizarre low whinnying sound ('drumming') by vibrating outer tail-feathers. **Status** Common breeding visitor in all suitable habitats and common passage migrant, late April to mid-September.

Swinhoe's Snipe *Gallinago megala* — 27–28 cm

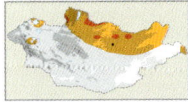

ID Nearly identical to Pin-tailed Snipe and probably not safely distinguished from that species in field, but slightly larger and longer-tailed, more yellowish, and plumage a little less 'contrasting' than Common or Pin-tailed Snipes; lacks white trailing edge to secondaries and underwings (cf. Common Snipe). See distribution of similar snipes. **Voice** Similar to Common Snipe but higher-pitched, and notes of display song more drawn-out. **Habitat** Breeds in tall vegetated areas of swamps and damp forest clearings; uses a variety of wetlands on migration. **Behaviour** Similar to Common Snipe. **Status** Rare breeding visitor and uncommon passage migrant to northern Mongolia from Hövsgöl east to Dornod province, late April to early September.

Pin-tailed Snipe *Gallinago stenura* — 25 cm

ID Very similar to Swinhoe's and Common Snipes, but in flight lacks the latter's white trailing edge to secondaries, and legs project far beyond short tail. On ground, look for broad ('bulging') pale supercilium combined with narrow dark (loral) eye-stripe at base of bill (cf. Common Snipe). Diagnostic pin-shaped outer tail-feathers rarely visible while landing. **Voice** Flight call similar to Common Snipe but a little softer. **Habitat** Nests in boggy mountain tundra and flooded taiga forest, as well as wet meadows and marshes; may be found in almost any wetland on migration. **Behaviour** Similar to Common Snipe. **Status** Uncommon breeding visitor across forest and steppe regions and uncommon passage migrant throughout Mongolia, including Gobi, late April to mid-September.

Solitary Snipe *Gallinago solitaria* — 30 cm

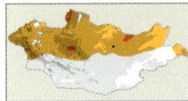

ID Readily distinguished from other Mongolian snipe species by large size; darker plumage; chunky form; gingery-brown wash to breast and hind-neck; lack of white trailing edge to secondaries; whitish (vs. cream-coloured) face pattern and mantle stripes; and slower, heavier flight. **Voice** Similar to Common Snipe, but harsher, *kensh*. **Habitat** Nests in bogs above tree line as well as brushy edges of cold mountain streams and springs; uses other wetland types on migration. **Behaviour** Less gregarious than other snipes. Slowly bobs or rocks body while feeding. **Status** Rare and local breeding visitor and passage migrant across northern Mongolia (from the Mongol-Altai mountain range to Dornod province), late April to mid-September; very rare winter visitor in the west.

Eurasian Woodcock *Scolopax rusticola* — 33–38 cm

ID A long-billed, forest-nesting wader, superficially resembling the *Gallinago* snipes, but readily distinguished by heavy, bulky form, rich chestnut-brown coloration above, and prominent head bars (vs. stripes). **Voice** Male in display makes a series of growling grunts (*quort, quort, quor-ort*), followed by a sharp *chiwich*. When flushed, alarm call is *chiki-chiki-chiki*. **Habitat** Moist forest with openings adjacent to wet areas. On passage found in thickets and even wet areas in steppe or desert. **Behaviour** Active mainly at dawn and dusk, feeding mostly at night. Shy and skulking on breeding grounds. Flight slow and straight on deeply bowed wings (compare snipe species). Male in display flight ('roding') flies above treetops with slow, deliberate wing-beats, with bill pointing down, calling regularly. **Status** Uncommon breeding visitor, mainly to northern forested areas, and fairly common passage migrant, including southern Gobi, late April to mid-September.

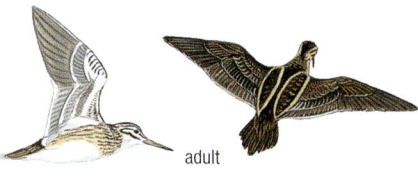

adult

adult

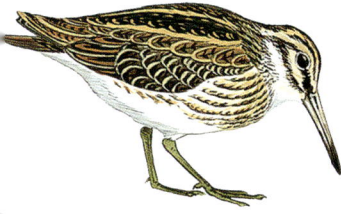

Common Snipe

Jack Snipe

adult

adult

adult

adult

Swinhoe's Snipe

Pin-tailed Snipe

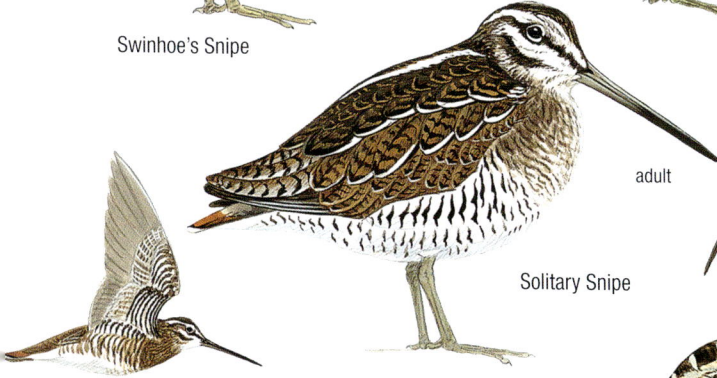

adult

Solitary Snipe

adult

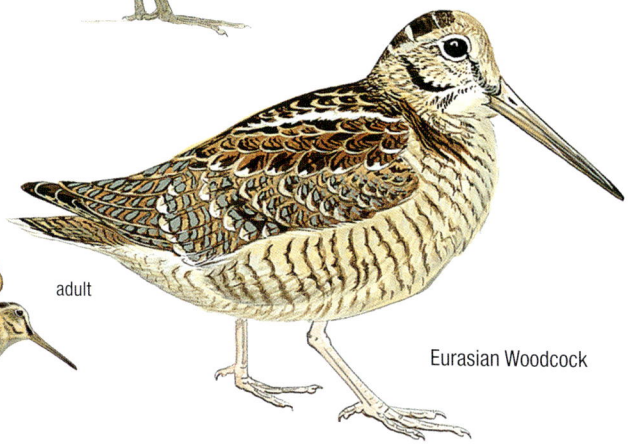

Eurasian Woodcock

PLATE 37: GODWITS, DOWITCHERS & PRATINCOLE

Black-tailed Godwit *Limosa limosa*

36–44 cm

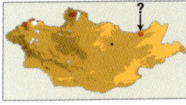

ID In breeding plumage, Black-tailed is best distinguished from Bar-tailed by largely white and barred lower breast and belly (vs. rufous and unbarred); extensive orange base to longer, straighter (vs. thinner, more strongly upturned) bill; and notably longer legs. Juv Black-tailed is largely smooth tawny with no streaking on head, neck and breast, and coarse spotting on mantle. Non-breeding adult is grey overall. Both juv and adult winter Bar-tailed are heavily, but finely, streaked on head, neck, breast and mantle. In flight, Black-tailed differs from Bar-tailed by highly distinctive white wing-stripes, wide black terminal tail-band and legs that extend well beyond tail. Compare also Asian Dowitcher. **Voice** A high-pitched and emphatic yapping *ke-ke-ke* or *kek-kek-kek*. **Habitat** Nests in wet meadows, shallow marshes, bogs and lake margins. Occurs in other open wetland habitats where waders gather on migration, including lakes in arid steppe and Gobi Desert. **Behaviour** Wades into deep water to feed, and probes in soft sediments on shoreline. Highly gregarious on migration. **Status** Rare and local breeding visitor and fairly common passage migrant throughout, late April to mid-August. **Conservation** Considered Near Threatened globally due to rapid decline in much of its population.

Bar-tailed Godwit *Limosa lapponica*

37–41 cm

ID For comparison with Black-tailed Godwit, see account for that species. In Mongolia, Bar-tailed Godwit is most easily confused with more common Asian Dowitcher. Latter is notably smaller, with shorter, dark (not bicoloured) bill that is thicker at base and comparatively blunt-tipped. **Voice** Similar to Black-tailed Godwit, but slower and lower in pitch. **Habitat** Nests in tundra and boggy taiga in Arctic; frequents a variety of shores and other wetland types on migration. **Behaviour** With shorter legs, less inclined to wade deeply than Black-tailed Godwit. **Status** Very rare passage migrant in appropriate habitat, late April to early May and mid to late August. Note: This species holds the record for the longest non-stop flight of any bird species, migrating 11,500 km in nine days from Alaskan breeding grounds to wintering grounds in New Zealand.

Long-billed Dowitcher *Limnodromus scolopaceus*

24–30 cm

ID A rare, snipe-like wader, most likely to be confused with Asian Dowitcher, but smaller with a distinctive face pattern (pale supercilium), yellowish-green legs and bill, and prominent streaking or scaling on head, neck, breast and flanks. In flight, shows narrow white trailing edge to wing and white wedge on lower back. **Voice** Call a single, high, sharp keek, or repeated chattering *keek-keek* or *kyik-kyik-kyik*. **Habitat** Muddy wetlands. **Behaviour** Feeds mainly by probing in damp ground. **Status** Vagrant. One record from valley of Degee River, Dornod province, 22 July 1977. Possibly migrates through eastern Mongolia, late April to early June and mid to late August.

Asian Dowitcher *Limnodromus semipalmatus*

33–36 cm

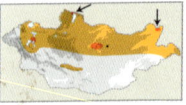

ID Superficially similar in appearance in all plumages to Bar-tailed Godwit, but with broad-based, straight (vs. upturned), all-black (vs. bicoloured) bill with swollen (vs. pointed) tip. Tends to tilt head downwards rather than more horizontally, and feeds with rapid 'sewing machine' motion rather than more deliberate probing and lifting. Vagrant Long-billed Dowitcher is notably smaller, with greenish (vs. blackish) legs, more complex face markings and prominent streaking and scaling in breeding plumage. **Voice** Calls rendered variously as *tye-chew, kiaow, chet chet* and *chowp*; when nesting, gives a soft *kerwick* or *kru-ku*. **Habitat** Breeds in small, loose colonies in wet meadows and grassy bogs, and uses other typical wader habitats, including lake shores in desert steppe and Gobi on migration. **Behaviour** Often associates on migration with Bar-tailed Godwit. **Status** Rare and local breeding visitor and passage migrant across most of the northern third of Mongolia and in Valley of the Lakes, late April to early September. **Conservation** Considered Near Threatened globally due to small total population and destruction of wintering habitat. Listed as Near Threatened in the Mongolian Red Book (2013). Included in the Threatened Birds of Asia (2001). [Alt: Asiatic Dowitcher]

Oriental Pratincole *Glareola maldivarum*

23–24 cm

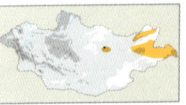

ID A medium-sized, coffee-coloured tern-like shorebird with black-bordered, yellowish throat patch, unlikely to be confused with any other Mongolian species. Juv is darker brown than adult above and below, with narrow pale edges to most feathers giving a scaly appearance. In all plumages, shows reddish-brown underwing-coverts and square white rump in flight. The very similar Collared Pratincole *G. pratincola* – recently recorded as a vagrant in southern Mongolia (see p. 268) – is distinguished by white trailing edge to secondaries, adult with more pronounced fork to tail and wing-tips not extending beyond tip of tail at rest. **Voice** Call described as 'a sharp *kyik, chik-chik* or *chet* often given in flight, also *ter-ack*, and a rising *trooeet*'. **Habitat** Nests colonially on dry, stony open steppe near lakes and rivers. On migration, roosts on sandy shores and other open habitats. **Behaviour** Very graceful in flight. Feeds mainly by hawking flying insects on the wing in swallow-like manner; often feeds and roosts in small flocks. **Status** Very rare passage migrant, mainly to eastern Mongolia, and possible, but unproven, rare and local breeding visitor; late April to late August.

non-breeding

non-breeding

breeding

melanuroides

breeding

limosa

non-breeding

juvenile

Black-tailed Godwit

♀
non-breeding

Bar-tailed Godwit

non-breeding

non-breeding

juvenile

♂
breeding

juvenile

breeding

Long-billed Dowitcher

non-breeding

juvenile

non-breeding

adult

breeding

juvenile

Asian Dowitcher

Oriental Pratincole

Little Curlew *Numenius minutus*   28–32 cm

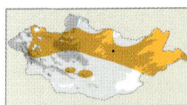

ID By far the smallest of the Mongolian curlews (not much larger than Pacific Golden Plover) with buffy plumage, and relatively short, not strongly decurved bill compared to other curlews. Head striping recalls Whimbrel, but has a broader, buffier supercilium. **Voice** Gives a rapid chattering *tititi* or *pepepe* in flight. **Habitat** In Mongolia, typically found in short grassland, including dry steppe and steppe lakes with shallow water. **Behaviour** Gregarious. Often feeds in uplands, but also wades and probes. **Status** Fairly common passage migrant in northern and eastern Mongolia, especially on eastern plains along valleys of Ulz, Herlen, Onon and Balj Rivers, where flocks of up to 300 have been recorded; late April to early May and early to late August.

Eurasian Curlew *Numenius arquata*   50–60 cm

ID This species is readily distinguished from Whimbrel and Little Curlew by its lack of bold head striping. Very similar to Far Eastern Curlew, but in flight shows prominent white wedge on rump and back and whitish (vs. brown-barred) vent and underwings. **Voice** A distinctive mournful *coor-leee*. Song contains similar notes, accelerating into bubbling trill. **Habitat** Nests in grassy or boggy areas, including wet meadows and forest clearings, but also forages in open uplands, including agricultural fields and open dry habitats near wetlands; gathers in other wetlands where waders congregate on migration. **Behaviour** Probes damp ground for food. Gregarious. **Status** Uncommon and local breeding visitor and uncommon but widespread passage migrant across northern Mongolia as well as in Valley of the Lakes, late April to late August. **Conservation** Considered Near Threatened globally due to recent sharp population declines.

Far Eastern Curlew *Numenius madagascariensis*   53–66 cm

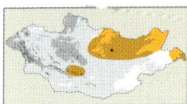

ID Best distinguished from very similar Eurasian Curlew in flight, when the latter shows prominent white rump and back and whitish (not brown-barred) underwings and vent. This species is also slightly larger, longer billed and warmer brown overall. **Voice** Similar to Eurasian Curlew but lower-pitched and more drawn-out. **Habitat** Various wetland types as well as uplands. **Behaviour** Similar to Eurasian and Little Curlews, with which it often congregates. **Status** Rare passage migrant, late April to early May and mid to late August, mainly to northern and eastern Mongolia; also recorded from Valley of the Lakes. **Conservation** Considered Endangered globally due to recent sharp population declines resulting from severe habitat destruction and deterioration.

Whimbrel *Numenius phaeopus*   40–46 cm

ID Bold head striping distinguishes this species from Eurasian and Far Eastern Curlews; Little Curlew is much smaller with overall buffy (not greyish) coloration. Two races occur in Mongolia: *N. p. phaeopus* has white back and rump, and *N. p. variegatus* has brown back and barred white rump. **Voice** Flight call a staccato, seven-note whistle. **Habitat** Grassy uplands, banks of rivers and lakes. **Behaviour** Similar to Eurasian Curlew but faster-moving on ground and in flight. **Status** Rare passage migrant, mainly on river and lake shores in northern Mongolia, but also on lakes in arid steppe and Gobi Desert where other waders congregate; passage late April and late August.

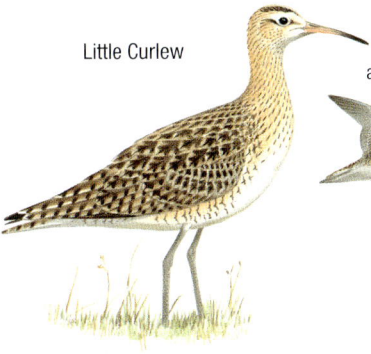
Little Curlew

adult

adult

Eurasian Curlew

adult

adult

Far Eastern Curlew

adult

adult

adult

Whimbrel

variegatus

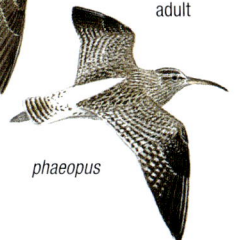

adult

phaeopus

Green Sandpiper *Tringa ochropus*   21–24 cm

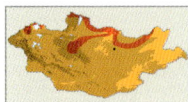

ID Similar to Wood Sandpiper, which it resembles in general form, behaviour and habitat preference, but distinguished by blackish (vs. brownish) coloration above; a prominent white eye-ring (vs. supercillium); shorter, green (vs. yellowish) legs; and in flight black (vs. whitish) under wings and broad (vs. narrow) tail barring. Both Green and Wood are significantly smaller, shorter-billed, and shorter-legged than Common Greenshank or Marsh Sandpiper. **Voice** When flushed gives a sharp 3-note *kleet veet veet*. Alarm call an abrupt *kwik kwik kwik*. **Habitat** Breeds in forested habitats near open water. Typically occurs along lake shores with emergent vegetation, small temporary ponds and sedgy ditches on migration, but also in other wetland types, e.g. oases in arid steppe and Gobi Desert. **Behaviour** Uses abandoned songbird (especially thrush) nests rather than building its own. Not very gregarious. **Status** Uncommon breeding visitor and passage migrant throughout Mongolia, late April to mid-September.

Wood Sandpiper *Tringa glareola*   19–23 cm

ID Similar to Green Sandpiper, which see for distinctions. **Voice** High-pitched *chiff iff iff* call given in flight and when flushed. **Habitat** Nests in thick vegetation on ground in taiga wetlands. On migration, occurs along riverbanks and lake shores with emergent vegetation, and marsh edges; also in various wetlands in arid steppe and Gobi Desert. **Behaviour** Feeds energetically, probing mud or wading and snapping prey from surface. Like Green Sandpiper, rather shy, often flushing suddenly and calling. **Status** Uncommon and local breeding visitor and common passage migrant throughout Mongolia, late April to early September.

Common Sandpiper *Actitis hypoleucos*   19–21 cm

ID Smaller, with much shorter legs and distinctly olive-brown coloration above compared to Green and Wood Sandpipers and tattlers. Lacks white rump, but has distinctive white 'gap' between breast side and folded wing, and shows white wing-stripe in flight. Possibly its most distinctive traits are behavioural (see below). **Voice** A loud shrill *swhee-swee-swee* when flushed and in flight; also a single longer *tweeeeh*. **Habitat** Uses wide variety of wetland types, showing some preference for rocky streams and lake shores. **Behaviour** Constantly bobs rear end of body and has a flight pattern of rapid, shallow wing-beats, interspersed with short glides, unlike that of any other wader in the region. Feeds mainly on shoreline. **Status** Common breeding visitor and passage migrant throughout country (absent only from highest elevations), late April to early September. **Taxonomy** Sometimes placed in genus *Tringa*.

Terek Sandpiper *Xenus cinereus*   22–25 cm

ID Upturned bill and greyish coloration above like Common Greenshank, but this species has much shorter orange-yellow (vs. grey-green) legs. In flight, lacks latter's white rump and back and shows a distinctive white trailing edge to secondaries. See also behaviour, below. **Voice** Common calls include a whistled *vee-vee-vee* and more rattling *wicka-wicka-wick*. **Habitat** Prefers lake shores and riverbanks, but also found in other wetland types. **Behaviour** Chases insects along shorelines in distinctive, hyperactive dashing manner. **Status** Uncommon to rare passage migrant throughout Mongolia, late April to early May and mid-August. **Taxonomy** Sometimes placed in genus *Tringa*.

Red-necked Phalarope *Phalaropus lobatus*   18–19 cm

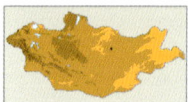

ID Breeding adults unmistakable: lead-grey above with wide rufous collar and tan streaking on mantle; females brighter than males. Juv and non-breeding birds best told from very similar Red Phalarope by lighter build; longer, needle-like (vs. deeper and blunter) bill; and darker, more streaked mantle. Juvs have brownish cast. **Voice** Flight call is a brief, chirruped *chep* or *kerrek*. **Habitat** Open water of lakes and ponds throughout much of the country. **Behaviour** Swims on surface, frequently spinning and pecking quickly all around for insect larvae and other small aquatic prey. **Status** Uncommon, though occasionally numerous, passage migrant, late April to early May and mid to late August.

Red Phalarope *Phalaropus fulicarius*   20–22 cm

ID Breeding adult unmistakable bright brick-red with white cheeks and black-tipped yellow bill. Non-breeding plumage best distinguished from very similar and more common Red-necked Phalarope by plain pale grey, unstreaked back and a thicker (vs. needle-like) bill. **Voice** A high-pitched, abrupt *tik* or *wit* in flight. **Habitat** Breeds in Arctic tundra; otherwise typically inhabits the open ocean or large lakes. **Behaviour** As Red-necked Phalarope, but feeds less energetically than that species. **Status** Vagrant. A first-year bird in Zulganai oasis, Ömnögobi province, 30 August and 2 September 2004 and one individual Lake Galuut, Dornod province, 17 July 2009. [Alt: Grey Phalarope]

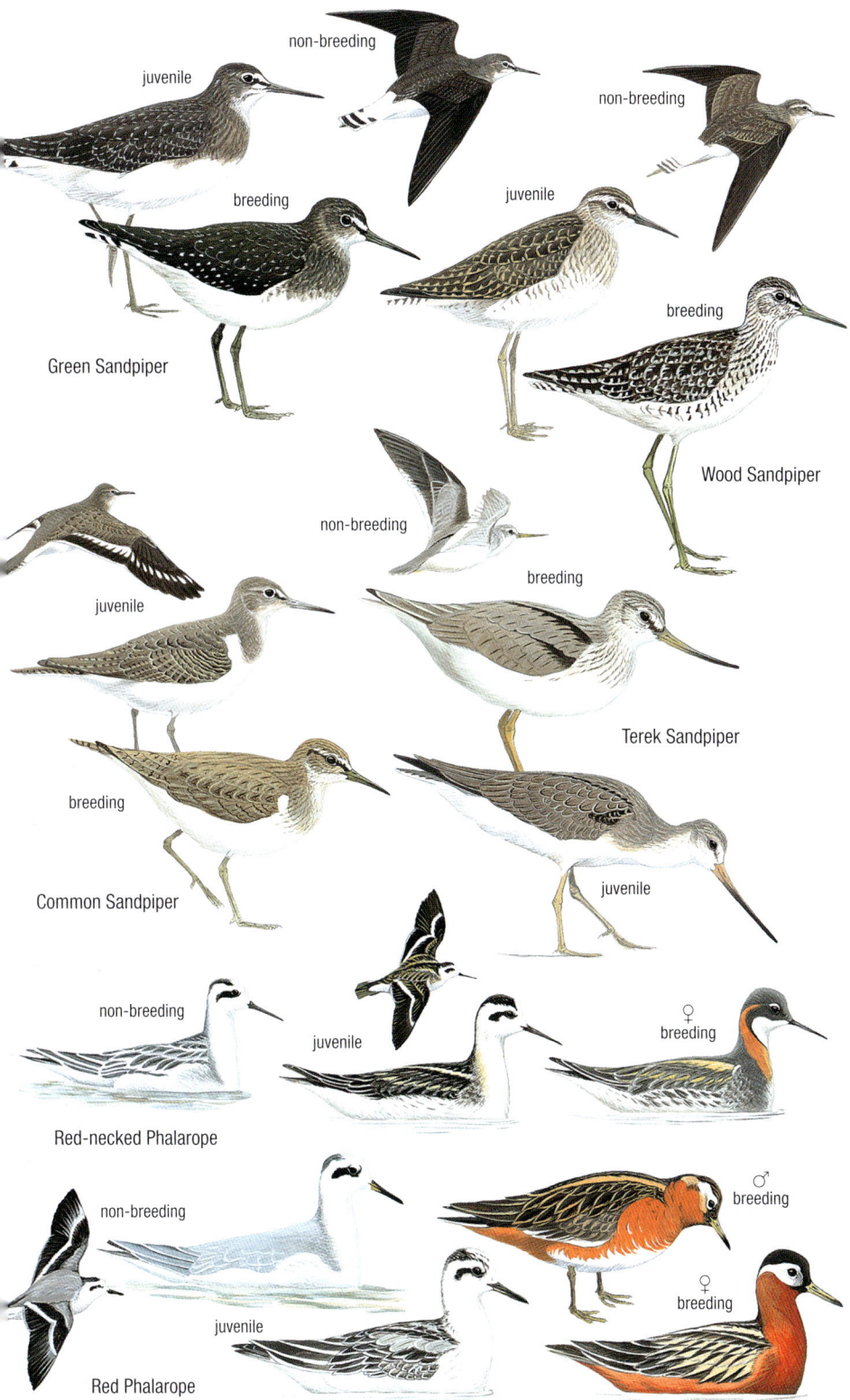

juvenile

non-breeding

non-breeding

breeding

juvenile

breeding

Green Sandpiper

Wood Sandpiper

non-breeding

breeding

juvenile

Terek Sandpiper

breeding

Common Sandpiper

juvenile

non-breeding

juvenile

♀
breeding

Red-necked Phalarope

non-breeding

♂
breeding

juvenile

♀
breeding

Red Phalarope

Common Greenshank *Tringa nebularia*   30–35 cm

ID Easily confused with Marsh Sandpiper, with which it shares general greyish coloration and notably long, grey-green legs. However, it is larger and stockier with relatively stout, slightly upturned (vs. very thin and straight) bill, and toes do not project very far beyond tail in flight. **Voice** Trisyllabic *tew tew tew* given in flight and on the ground. **Habitat** Margins of various aquatic habitats, including lakes and wetlands in desert steppe and Gobi Desert. **Behaviour** Wades, fishes and probes in mud, often with other waders and small flocks of its own species. **Status** Uncommon passage migrant throughout Mongolia, late April to early May and mid to late August.

Marsh Sandpiper *Tringa stagnatilis*   22–26 cm

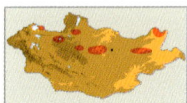

ID A smaller, more slender and delicate version of Common Greenshank with a shorter, thinner and straight (vs. slightly upturned) bill. Also proportionately long-legged with toes projecting well beyond the tail in flight. Differs from Green and Wood Sandpipers by much longer legs and bill; and in flight by long, wedge-shaped (vs. square) white patch on rump and back. **Voice** A shrill, single *piyoo* or *pyu*, sometimes repeated. Abrupt *iyup* when flushed. **Habitat** Nests in grassy marshes near lakes and ponds and in open bogs. **Behaviour** Like other *Tringa* waders, feeds in water and on shorelines in various aquatic habitats on migration. **Status** Uncommon and local breeding visitor and fairly common passage migrant throughout Mongolia, late April to late August.

Common Redshank *Tringa totanus*   27–29 cm

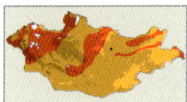

ID Compared to Spotted Redshank, with which it shares bright red legs and bill base, this species is smaller and more compact with an overall brownish (vs. grey) coloration, a heavily streaked breast and a shorter bill. In flight, wings show diagnostic broad white trailing edges. Juv. with yellow-orange legs and bill base, might be confused with some Ruffs, but is readily distinguished by heavy streaking, straight bill and very different build. **Voice** A whistling *kew hu hu hu* or *te hoo hoo* frequently given in flight, especially when nesting; also a more drawn-out *tyeuuuuu*. **Habitat** Nests in wet meadows, grassy marshes and margins of rivers and lakes. On migration, occurs with other waders on lake shores and other wetlands in arid steppe and Gobi Desert. **Behaviour** Alert and noisy on breeding grounds. Quite gregarious when not breeding. **Status** Locally common breeding visitor and passage migrant throughout Mongolia, late April to late August (early September in the Gobi).

Spotted Redshank *Tringa erythropus*   29–32 cm

ID Similar to Common Redshank (the only other long-legged wader with bright red legs and bill base). However, Spotted Redshank is much more slender and gracefully proportioned, with a much longer bill, which droops slightly at the tip. Breeding and winter plumage adults are also readily distinguished by general coloration (black or grey respectively vs. brownish) and lack of streaking. In flight, toes fully project beyond tail, and wings lack bold white trailing edges shown by Common Redshank. **Voice** A distinctive *tu-ick* flight call. **Habitat** Uses a wide variety of typical wader habitats on edges of wetlands and water bodies. **Behaviour** Frequently wades deeply and feeds energetically on aquatic invertebrates and small fish. **Status** Uncommon passage migrant throughout Mongolia, including oases in arid steppe and Gobi Desert, late April to early May and mid to late August (single birds to early September).

juvenile

Common Greenshank

non-breeding

breeding

non-breeding

juvenile

Marsh Sandpiper

breeding

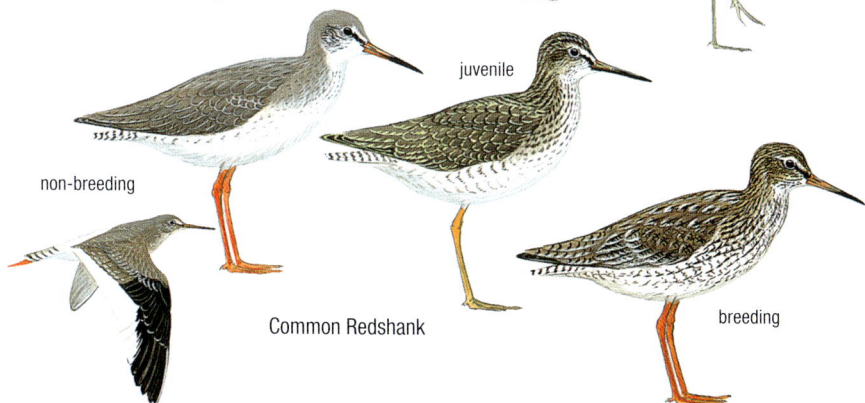

juvenile

non-breeding

Common Redshank

breeding

juvenile

non-breeding

Spotted Redshank

breeding

Grey-tailed Tattler *Tringa brevipes*

23–27 cm

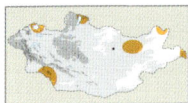

ID Breeding adult distinguished from all other Mongolian waders (except the following vagrant species, which see for distinctions) by combination of barred (vs. streaked) breast and flanks; comparatively short, yellow legs; and absence of white rump/back patch in flight. Smooth grey upperparts in addition to previous two characters distinguish juvs and autumn adults. Common Sandpiper is much smaller, browner and unbarred; Terek Sandpiper has distinctly upturned (vs. straight) bill. **Voice** A Grey Plover-like double whistle, *tyuu-deee*. **Habitat** Prefers rocky, sandy or muddy shores of rivers and lakes, but may occur in other wetland habitats on migration. **Behaviour** Typically solitary (rarely in small flocks). Feeds on shore but also wades in shallows. **Status** Rare passage migrant, with single records from lakes and river valleys across north and east, including Hövsgöl and Tari lakes, Herlen River and several localities in Trans-Altai Gobi; passage probably late April to mid-May and mid to late August. **Conservation** Considered Near Threatened globally due to moderately rapid population decline resulting from habitat loss and disturbance, as well as hunting. **Taxonomy** Formerly placed in genus *Heteroscelus*. [Alt: Grey-rumped Tattler, Polynesian Tattler]

Wandering Tattler *Tringa incana*

26–29 cm

ID Very similar to Grey-tailed Tattler, but darker. Breeding birds have heavier, more extensive barring on underparts, including across centre of belly and vent. At rest, longer primary tips extend beyond tip of tail. Non-breeding birds almost inseparable from Grey-tailed in field except by call. **Voice** Flight call a series of short, clear, piping notes at constant pitch *pi pi pi pi*. **Habitat** Favours rocky shores. **Behaviour** As Grey-tailed Tattler. **Status** Vagrant. Two records: an adult on western shore of Lake Hövsgöl, 25 July 1903 and several birds on Khoroo and Jargalant Rivers and at Lake Hövsgöl, Hövsgöl province, mid-August 1972. **Taxonomy** Formerly placed in genus *Heteroscelus*.

Ruff *Calidris pugnax*

M 26–32, F 20–25 cm

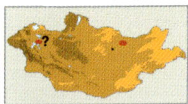

ID Male in breeding plumage unmistakable with variations of a multicoloured 'ruff' of feathers on head, neck and breast. Breeding females, non-breeding adults and young birds come in confusing array of sizes and plumage patterns, but with distinctive form that combines small head, long neck, plump belly, medium-length decurved bill and longish pale orange to bright orange legs. In flight, white oval patches on sides of tail and thin white wing-stripes above are diagnostic. **Voice** Generally silent, even in display, except for short grunts, *kuk* or *kurr*. Flight call is a shrill *hoowhee*. **Habitat** Breeds in freshwater marshes and wet grasslands but occurs on shores and other wetland wader habitats, including desert steppe and Gobi on migration. **Behaviour** Males gather in numbers and display communally for females on lekking grounds. **Status** Rare and very local breeding visitor at Lake Airag (Great Lakes Depression) and upper Tuul and Herlen Rivers (Hentii mountain range) and uncommon passage migrant throughout, late April to late August (early September in the Gobi). **Taxonomy** Formerly placed in genus *Philomachus*.

Red Knot *Calidris canutus*

23–25 cm

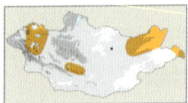

ID In breeding plumage might be confused with Curlew Sandpiper, but readily distinguished by notably stout build, rufous (vs. chestnut) coloration, shorter dark greenish legs, and short, straight bill. Non-breeding birds slightly resemble non-breeding Grey Plover or Sanderling due to stocky form, but are easily separated by leg colour and finer bill. **Voice** In flight, gives a low-pitched staccato *knutt knutt knutt*. **Habitat** Mainly lake shores. **Behaviour** Feeds on open beaches and mudflats, often with other waders. **Status** Rare passage migrant through the Great Lakes Depression, Valley of the Lakes and eastern Mongolia, late April to early May and early to mid-August. **Conservation** Considered Near Threatened globally.

Sanderling *Calidris alba*

20–21 cm

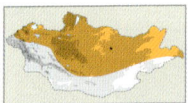

ID In breeding plumage could be confused with Red-necked or Little Stints but is larger and stockier with a heavier, blunter bill and lacks hind toe. Non-breeding birds are strikingly white overall, juv with black-and-white scaling on mantle; adult winter uniform pale grey above. In flight, shows broad white wing-stripe. **Voice** A short, hard *cheet cheet* in flight; also a sharper and higher *pleeet* or *veek*. **Habitat** Prefers sandy shores and open beaches. **Behaviour** Gregarious. Runs hyperactively along shoreline, chasing mobile invertebrate prey. **Status** Rare passage migrant across Mongolia, south to Valley of the Lakes and east to Dornod province, late April to early May and late August to early September.

juvenile

non-breeding

breeding

Grey-tailed Tattler

non-breeding

Wandering Tattler

♂
non-breeding

♂
non-breeding

♀
breeding

♂
breeding

♂
display

juvenile

Ruff

non-breeding

non-breeding

juvenile

juvenile

non-breeding

Red Knot

breeding

Sanderling

breeding

non-breeding

Little Stint *Calidris minuta*   12–14 cm

ID Very similar to more common Red-necked Stint, with which it is likely to occur on migration. In breeding plumage, Little Stint has more streaks on face and upper breast, which gives a less bright tone to the rufous coloration the two species share. Except for winter plumage adults, Little Stint shows a prominent white V on mantle at rest and in flight. For additional distinctions, see Red-necked and other stints and Sanderling. **Voice** A sharp *chit* or *stit* and softer *pipipi* given in flight; very similar to Red-necked Stint. **Habitat** Typically, wet edges of lakes and rivers. **Behaviour** As Red-necked Stint. **Status** Uncommon passage migrant in appropriate habitat throughout most of Mongolia, including arid steppe and Gobi Desert, late April to early May and early to late August.

Red-necked Stint *Calidris ruficollis*   13–16 cm

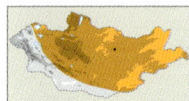

ID Very similar to Little Stint and may be indistinguishable in some plumages and conditions. Best points of distinction for Red-necked are unstreaked rufous-orange throat, foreneck and breast (breeding plumage only); slightly thicker bill, squarer head and shorter legs; and longer primary projection creating slightly more elongated appearance. Lacks (or has very weak) prominent mantle V of Little Stint. Adult non-breeding plumage almost identical to Little Stint, but slightly cleaner and greyer above. See also other stint species and Sanderling. **Voice** Very similar to Little Stint; perhaps slightly 'coarser'. **Habitat** Lake shores and other wetland edges where waders gather on migration. **Behaviour** Pecks and probes actively at water's edge. Commonly flocks with other waders. **Status** Common passage migrant throughout most of Mongolia, including arid steppe and Gobi Desert, late April to early May and early to late August. **Conservation** Considered Near Threatened globally. [Alt: Rufous-necked Stint]

Long-toed Stint *Calidris subminuta*   13–16 cm

ID Readily distinguished from Red-necked and Little Stints by pale yellowish-greenish legs, which project beyond tip of tail in flight; more delicate build with finer slightly decurved bill; and prominent supercilium. Often looks longer-necked in the field. In breeding plumage, lacks rufous face, neck and breast. Compare also Temminck's Stint and Sharp-tailed Sandpiper. In winter, upperparts more heavily marked than Little Stint. **Voice** A purring, quiet *chrrip* or *prrrrrp* in flight. **Habitat** Breeds in grassy boreal marshes. On migration, typically feeds on grassy edges of lakes, including those in arid steppe and Gobi Desert. **Behaviour** Often feeds in cover of bordering vegetation, pecking and probing for invertebrates. **Status** Probably nests locally in northern Mongolia, though breeding not confirmed. Fairly common passage migrant, late April to early May and early to late August.

Temminck's Stint *Calidris temminckii*   13–15 cm

ID Compared to other stints, Temminck's has uniform drab greyish-brown appearance overall with no striking markings. Rather squat and short-legged. Pale greenish-grey legs are duller than those of Long-toed Stint, which is strikingly different in posture and has strong supercilium and prominently streaked breast. In flight, shows diagnostic white sides to dark-centred rump and tail. **Voice** Flight call is a trilling *tirrrrr* or *tirrrr-rrrr-rrr*. **Habitat** Wetland shores and marsh edges, preferring to feed in sparse vegetation rather than on open flats or beaches. **Behaviour** More deliberate in its movements than other stints with a characteristic creeping motion. Occurs in small flocks, often with other stints. **Status** Common passage migrant throughout Mongolia, late April to early May and mid-August to early September.

juvenile

non-breeding

breeding

Little Stint

non-breeding

juvenile

breeding

Red-necked Stint

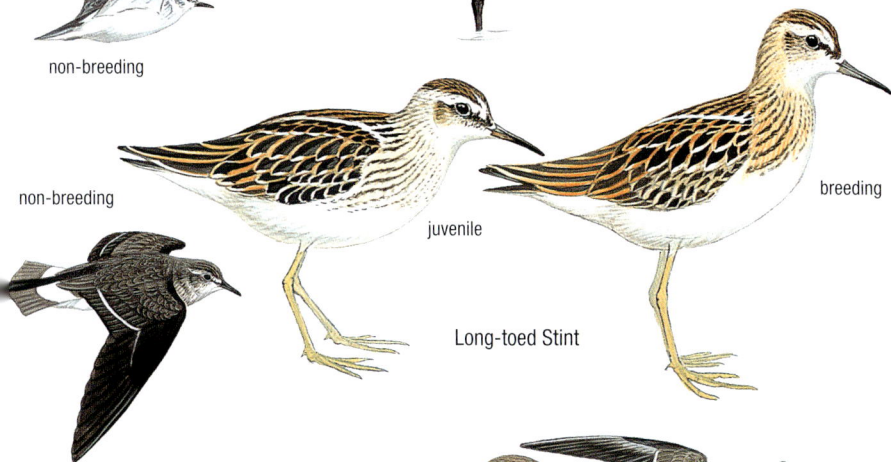

non-breeding

non-breeding

juvenile

breeding

Long-toed Stint

juvenile

breeding

non-breeding

Temminck's Stint

Curlew Sandpiper *Calidris ferruginea*   18–23 cm

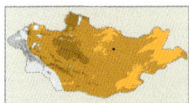

ID In breeding plumage, combination of dark chestnut-red coloration, long black decurved bill, and longish black legs is unique. (Red Knot is rufous with shorter bill and short, greenish legs). Non-breeding birds most likely to be confused with much rarer Dunlin and Broad-billed Sandpiper, but are longer-legged with more elegant, upright posture, paler head pattern and longer, more distinctly decurved bill with finer tip. In flight, white rump-patch diagnostic. **Voice** In flight, gives a soft purring *prrrrp* or *chrrrup*. **Habitat** A wide variety of open shores and wetland edges. **Behaviour** Often wades belly-deep as it probes in bottom sediments for prey. Forms flocks with other waders. **Status** Common passage migrant across Mongolia, including lakes in arid steppe and Gobi Desert, late April to early May and early to late August. **Conservation** Considered Near Threatened globally.

Dunlin *Calidris alpina*   16–22 cm

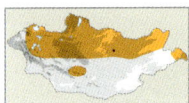

ID Bright rufous back and crown and black belly patch of breeding plumage are unique to this species. Non-breeding birds may be confused with more common Curlew Sandpiper, but are shorter-legged, 'dumpier' in stature and have a slightly shorter, less pointed bill than the latter. In flight, lacks distinctive white rump of Curlew Sandpiper. **Voice** Flight call is a harsh, buzzy whistle *skrreeer* or *schreet*. **Habitat** A variety of open shores and wetland edges. **Behaviour** Highly gregarious on migration. Probes for prey mainly at water's edge, wading less than the longer-legged Curlew Sandpiper. **Status** Relatively rare passage migrant across northern Mongolia including the far east, late April to early May and mid to late August.

Sharp-tailed Sandpiper *Calidris acuminata*   17–22 cm

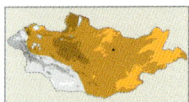

ID Resembles large Long-toed Stint with rufous crown and feather edgings above, a prominent white supercilium and greenish-yellow legs. Breeding adult shows distinctive arrow-headed streaks on flanks. Often appears long-necked and small-headed. Juv has diagnostic bright orange-buff breast with only light streaking. Compare Pectoral Sandpiper. **Voice** Flight call is a soft *trrrrt* or *purrrri*. **Habitat** Moist grasslands and drier edges of lake shores. **Behaviour** Prefers to forage in thin vegetation rather than in open. Gregarious with other waders. Often quite approachable. **Status** Common passage migrant across Mongolia, including far east, south to Valley of the Lakes, late April to early May and mid-August to early September.

Pectoral Sandpiper *Calidris melanotos*   19–23 cm

ID A fairly large sandpiper, likely to be confused only with Sharp-tailed Sandpiper, which it resembles in general form. However, it lacks bright rufous tones and arrow-headed streaks on flanks and in all plumages shows a diagnostic sharp cut-off between a heavily streaked upper breast and white belly. **Voice** When flushed, gives a rough *trrit* or *krrit* similar to that of Curlew Sandpiper. **Habitat** Like Sharp-tailed Sandpiper. **Behaviour** Similar to Sharp-tailed Sandpiper. **Status** Vagrant. A single bird photographed at Lake Tsagaan in Norovlin district, Hentii province, 30 August 2010.

Broad-billed Sandpiper *Calidris falcinellus*   16–18 cm

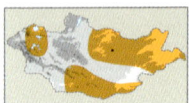

ID In autumn, might be mistaken for juv Dunlin or Curlew Sandpiper, but is slightly smaller and notably shorter-legged with bill 'kinked' downward at tip rather than more evenly decurved. Prominently striped head is diagnostic in all plumages, though more obscure in juv and non-breeding adult. **Voice** A dry, rolling Sand Martin-like *chrrreep* or *trrrrr trrrr trrrrk* in flight. **Habitat** Favours muddy flats and boggy edges of rivers and lakes; flocks with other waders **Behaviour** Moves slowly when foraging, probing deliberately rather than rapidly like stints. Often tame and approachable. **Status** Fairly common passage migrant in Great Lakes Depression and river valleys in Hentii range through Dornod province; also in lakes of arid steppe and Gobi, late April to early May and mid- to late August. **Taxonomy** Formerly placed in genus *Limicola*.

juvenile

juvenile

non-breeding

breeding

Curlew Sandpiper

juvenile

juvenile

non-breeding

Dunlin

breeding

juvenile

breeding

non-breeding

Sharp-tailed Sandpiper

juvenile

juvenile

breeding

non-breeding

Pectoral Sandpiper

Broad-billed Sandpiper

Pomarine Skua *Stercorarius pomarinus*

46–51 cm

ID Skuas are gull-like, but are darker than even juv gulls. They occur in two colour morphs. In adults the dark morph appears uniformly chocolate-brown; the light morph has contrasting white underparts, throat and nape, and yellowish cheeks. In breeding plumage, Pomarine may be readily distinguished from Arctic Skua by twisted, blunt-ended central tail feathers. In other plumages, distinctions are comparative: Pomarine is larger, 'chestier' bird with broader wing bases; is notably dark (almost blackish) brown compared to Arctic's paler plumage tones; tends to have very prominent dark breast-band and conspicuous dark barring on flanks; has a heavier bill with pink base and black tip (vs. Arctic's uniformly dark); has more ponderous, less agile manner of flight; and tends to show more white in 'flashes' at base of primaries. Juv Arctic tends to have cinnamon tinge to head and underparts and more conspicuous barring below, while blue-grey bill base in juv Pomarine is often helpful distinction, even at long range. These birds are highly variable in plumage, however, and colour tones vary with ambient light conditions; therefore many skuas cannot be identified to species even by experienced observers. **Voice** Usually silent away from breeding grounds. **Habitat** Breeds on Arctic tundra; otherwise a highly pelagic seabird, except on migration when it may visit large inland lakes. **Behaviour** Piratical, stealing prey from other birds; also scavenges. **Status** Vagrant. One at Lake Khar, Khovd province (undated), one adult light morph at Dashinchilen Tsagaan Nuur, Bulgan province, 25–26 August 2006 and one at Lake Bööntsagaan, Bayankhongor province, 30 September 2006. [Alt: Pomarine Jaeger]

Arctic Skua *Stercorarius parasiticus*

41–46 cm

ID Smaller and slighter than Pomarine Skua in all plumages. For detailed distinctions, see above. **Voice** Usually silent away from breeding grounds. **Habitat** Breeds in Arctic and subarctic tundra; otherwise a highly pelagic seabird. Occasionally recorded on large inland lakes during migration. **Behaviour** Very similar to Pomarine Skua. **Status** Vagrant. Single birds near the Buur River, Selenge province, August 1991; and Lake Airag, Uvs province, August 2000. [Alt: Arctic Jaeger, Parasitic Jaeger]

Pallas's Sandgrouse *Syrrhaptes paradoxus*

30–41 cm

ID Unlikely to be confused with any other Mongolian species due to its unique form – resembling a hybrid between a grouse and a pigeon with long, pointed wings and 'pin-tail' – and its behaviour and habitat. **Voice** Various calls, including *quat*, harsh *kikiki* and crooning *cryoo-ryoo*, call habitually given in flight. **Habitat** Arid steppe and Gobi Desert. **Behaviour** Walks on ground, foraging for seeds and other plant material. At dawn and dusk, flies in flocks to waterholes where males soak their specially modified belly feathers to carry water to nestlings. Nest is scrape on open ground. Forms large post-breeding flocks and has been known to undertake mass dispersals, reaching as far as British Isles. **Status** Common resident breeder across dry, open regions with sandy soils, from Great Lakes Depression to Dornogobi. Locally and seasonally abundant when conditions induce irruptive migrations of wintering flocks.

Pomarine Skua

breeding
pale morph

breeding
dark morph

juvenile

Arctic Skua

breeding
dark morph

breeding
pale morph

juvenile

♂

Pallas's Sandgrouse

♀

♂

Pallas's Gull *Ichthyaetus ichthyaetus*   58–67 cm

ID Breeding adult readily distinguished from all other Mongolian gulls by combination of large size; yellow bill with black-and-red tip; and distinctive wing-tip pattern. Other plumages might be mistaken for winter or subadult Mongolian Gull or other large gulls, but Pallas's always shows dark facial mask with prominent white 'spectacles'; notably long, pinkish-yellow bill with solid black tip; dark smudge at base of neck; and, in flight, contrasting pale grey panel across middle of upperwings and clear white (unspeckled) tail with clean black tail-band. **Voice** Deep-voiced, almost crow-like. **Habitat** Lakes and large rivers. **Behaviour** Nests colonially near water. Scavenges and takes assorted live prey. **Status** Locally uncommon breeding visitor to Great Lakes Depression and rare passage migrant to suitable habitats in central and western Mongolia, late April to early September. **Taxonomy** Formerly placed in genus *Larus*. [Alt: Great Black-headed Gull]

Mongolian Gull *Larus mongolicus*   56–68 cm

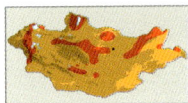

ID The most abundant and widespread gull in Mongolia. It resembles Mew Gull in some plumages, but is much larger with correspondingly large and differently marked bill in all plumages. This species is part of the Holarctic 'Herring Gull' complex, now split into various species or races, depending on taxonomic interpretation. The problem of distinguishing it from many similar large gulls is well illustrated by the variation among Mongolian Gulls themselves; adult leg colour varies on a spectrum from orange-yellow to pink; and iris colour ranges from pale yellow to dark. However, no other 'Herring Gull' type has (yet) been identified in Mongolia. **Voice** Typical harsh large-gull calls, *kaaa*, *kwaaa* and territorial bugling. **Habitat** Large lakes and rivers, typically nesting on islands, but also on shores and cliffs and in reedbeds. Also occurs on steppe, where it predates Brandt's Vole on migration. **Behaviour** Colonial and gregarious, frequently flocking with other large gulls. In addition to aquatic prey, feeds on small mammals, reptiles and amphibians, eggs and young of other waterbirds, including gulls, and carrion. **Status** Common to abundant breeding visitor and passage migrant throughout Mongolia, absent as a breeding species only in the most arid regions of the southern deserts and mountains, mid-April to early September. **Taxonomy** IOC treats this form as a subspecies of Vega Gull *L. vegae*.

Glaucous Gull *Larus hyperboreus*   64–77 cm

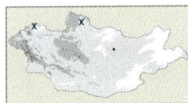

ID A large, stocky, very pale grey-mantled gull with white primaries. In Mongolia, most similar to Mongolian Gull, but easily distinguished in all plumages by total absence of black in wing-tips. Juv is pale tan (café-au-lait); 1st- and 2nd-winter birds can be nearly pure white. **Voice** Loud, throaty *guwaa* and *ag-ag-ag*. **Habitat** Large lakes and rivers. **Behaviour** Similar to Mongolian Gull. **Status** Vagrant. Single birds at delta of Khoroo River in Lake Hövsgöl, Hövsgöl province, 6 August 1971, and Lake Uvs, Uvs province, May 2010.

Pallas's Gull

non-breeding

non-breeding

1st-winter

breeding

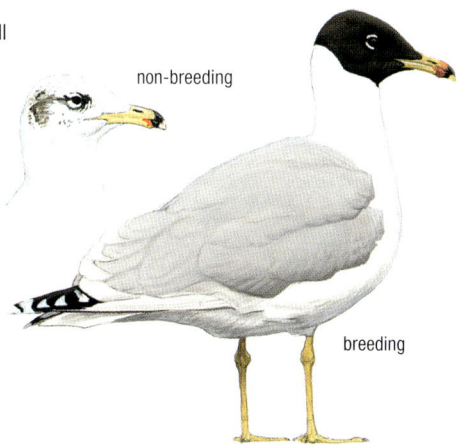

Mongolian Gull

1st-winter

1st-winter

breeding

non-breeding

1st-winter

juvenile/
1st-winter

non-breeding

Glaucous Gull

Relict Gull *Ichthyaetus relictus*   46 cm

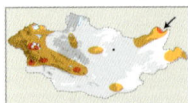

ID Adult most likely to be confused with Black-headed Gull, but readily distinguished by sooty-black (not chocolate brown) hood (in breeding plumage); heavier build; thicker, blunter bill; and in flight by distinctive wing-tip pattern. On ground, birds look small-headed with disproportionately bulky body. Adult winter, juv and subadult plumages resemble Mew Gull but head and bill shape are distinctive and Relict never shows tan streaking and barring of imm and winter Mew Gulls. **Voice** Chuckling *ka-ka-ka-kee-a*. **Habitat** Large lakes; also occurs in steppe on migration. **Behaviour** Nests colonially, mainly on lake islands where it is extremely sensitive to water levels relative to its nesting sites and vegetation cover, resulting in limited breeding success in many years. **Status** Rare and local breeding visitor to fresh and saline lakes in Great Lakes Depression, the Valley of the Lakes and Lake Tari (formerly Lake Tooroi), Dornod province, and uncommon passage migrant to many localities across Mongolia, late April to early September. Many migrant birds are presumed to be non-breeding wanderers unable to find suitable nesting conditions. **Conservation** One of the world's rarest gulls, recognised as a distinct species only in 1970. Nests only on desert lakes in Mongolia, Kazakhstan and China, and winters mainly in coastal China and Korea. Population declining. Considered Vulnerable globally. Listed as Threatened in the Mongolian Red Book (2013). **Taxonomy** Formerly placed in genus *Larus*.

Black-headed Gull *Chroicocephalus ridibundus*   35–39 cm

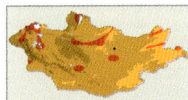

ID Most common 'hooded' gull in Mongolia. In breeding plumage, readily distinguished from other regularly occurring dark-headed gulls (despite its name!) by its chocolate brown head and white leading edge to wings. For comparisons with rarer gull species see Relict, Brown-headed, Slender-billed and Little Gulls. **Voice** Harsh drawn-out *skreeer* and sharper *kek* calls. **Habitat** Nests in reedbeds and other dense vegetation on lake and river shores and visits other aquatic habitats, including arid steppe and desert lakes on migration. **Behaviour** Nests in dense colonies. In addition to foraging in and near water, frequently forages for insects in flocks in steppe and agricultural land. **Status** Common breeding visitor across Mongolia and fairly common passage migrant, late April to late August (early September in the Gobi). **Taxonomy** Formerly placed in genus *Larus*. [Alt: Common Black-headed Gull]

Slender-billed Gull *Chroicocephalus genei*   42–44 cm

ID In all plumages most closely resembles Black-headed and Brown-headed Gulls, but readily distinguished in breeding plumage by completely white head and pale iris; otherwise best told by its distinctly slender neck, head and bill and by faint (vs. dark) ear-spot of imm. **Voice** Calls similar to Black-headed Gull but more rasping. **Habitat** Lakes and other aquatic habitats. **Behaviour** Gregarious; flocks with other gull species. **Status** Vagrant. A single record at Lake Uvs, Uvs province, 19 June 1977. **Taxonomy** Formerly placed in genus *Larus*.

Brown-headed Gull *Chroicocephalus brunnicephalus*   41–45 cm

ID Breeding adult very similar to Black-headed Gull, but slightly larger. In breeding plumage the head is paler shade of brown, especially around the bill, and iris is pale yellow, not dark. In all plumages has longer, heavier bill and broader, more rounded wings, with much more black in wing-tips than Black-headed. First-winter resembles non-breeding adult, but distinguished by dark brown wing-coverts and dull orange legs and bill. Compare also Slender-billed Gull. **Voice** Similar to Black-headed Gull but lower-pitched. **Habitat** In Mongolia, brackish lakes. **Behaviour** Similar to Black-headed Gull. **Status** Vagrant. A 2nd-year bird at Lake Bööntsagaan, Bayankhongor province, 7 June 2004 and three at western shore of Lake Uvs, Uvs province, 5 July 2004. **Taxonomy** Formerly placed in genus *Larus*.

Relict Gull

breeding

non-breeding

non-breeding

1st-winter

breeding

non-breeding

Black-headed Gull

non-breeding

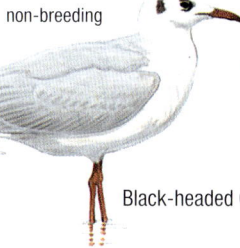

1st-winter

breeding

Slender-billed Gull

breeding

1st-winter

non-breeding

Brown-headed Gull

Little Gull *Hydrocoloeus minutus*   24–28 cm

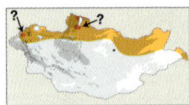

ID The world's smallest gull. In addition to size, adult is readily distinguished from other Mongolian gull species by black hood (in breeding plumage, vs. dark brown in Black-headed Gull); and, in flight, absence of any black on upperwings, and conspicuous blackish underwings with narrow white trailing edge. Juv and 1st-winter birds have striking black M-pattern spanning back and wings, similar only to immature Ross's Gull and Black-legged Kittiwake (both rare vagrants), which both lack Little Gull's dark crown patch and grey bar along trailing edge of secondaries in these plumages. **Voice** Tern-like, harsh *keck* or *keck-keck-keck*. **Habitat** Nests in reedbeds and on vegetated islands and shores of lakes and rivers, often in colonies of terns and other gull species. **Behaviour** Buoyant, agile and tern-like in flight; often catches insects in the air and picks food items from water's surface. **Status** Very rare and local breeding visitor; known to nest at only a few locations on Lake Achit in Great Lakes Depression and at Lake Hövsgöl; also an uncommon passage migrant across northern Mongolia, late April to late August. **Taxonomy** Formerly placed in genus *Larus*.

Mew Gull *Larus canus*   40–46 cm

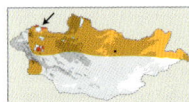

ID Medium-sized, white-headed gull. All plumages are broadly similar to those of Mongolian Gull, but the latter is 25–30% larger with much longer, heavier and differently marked bill. Size similar to Relict Gull, with juv and 1st-winter birds closest in plumage details, but differences in head and bill shape and colour, and much more extensive brown streaking and speckling on head, breast and mantle distinguish Mew Gull. Greenish leg colour also diagnostic in adult and 2nd-winter birds. See also Black-legged Kittiwake. **Voice** Calls a mewing *gya gya* or *gyu gyu*, also drawn-out *glyoooo*. **Habitat** Lakes and rivers. **Behaviour** Gregarious, usually colonial, often nesting with other gulls on islands. Defers to larger gulls at feeding gatherings. **Status** Very rare and local breeding visitor to a few lakes in Great Lakes Depression and possibly in suitable habitat in Khangai and elsewhere in northern Mongolia, and uncommon passage migrant throughout the rest of the country apart from most arid southern deserts and mountains, late April to early September. [Alt: Common Gull]

Black-legged Kittiwake *Rissa tridactyla*   38–40 cm

ID An agile, medium-sized gull with tern-like flight. Breeding adult distinguished from other gulls by combination of slender unmarked greenish-yellow bill, short black legs and 'ink-dipped' wing-tips with no white 'mirrors'. Non-breeding adult similar to breeding adult, but has dark grey bar on sides of head. In flight, juv and 1st-winter birds show bold black 'M' above, set off by clear grey mantle, black-tipped tail and white triangles in flight feathers; this pattern is superficially similar to Little Gull, but much 'cleaner'. Compare also Ross's Gull. **Voice** A high pitched tittering *kitti-wake*. **Habitat** Sea cliffs and open ocean; in Mongolia, lakes and other aquatic bodies. **Behaviour** Normally a highly pelagic gull, unlikely to linger long inland. **Status** Vagrant. One recorded in Herlen River, Töv province, on autumn migration (date unknown).

Ross's Gull *Rhodostethia rosea*   29–32 cm

ID Breeding adult unmistakable with black neck-ring, rosy blush below and red legs, in addition to diminutive size. Imm and non-breeding adult have black eye patch, creating an odd 'zombie' look. In flight, superficially resembles Little Gull but adult has medium-grey, not blackish underwings; bold mantle pattern of juv and 1st-winter birds is much 'cleaner' than in Little; and tail is quite long and pointed, not notched as in Little. Compare also Black-legged Kittiwake. **Voice** *Ku-wa ku-wa* call occasionally given in flight. **Habitat** Nests in Arctic tundra and winters in Arctic Ocean. **Behaviour** Graceful and tern-like in flight. Often forages on mud like a wader. **Status** Vagrant. A single adult on Lake Uvs, Uvs province, 9 June 1977.

breeding

non-breeding

breeding

juvenile

Little Gull

breeding

breeding

non-breeding

1st-winter

Mew Gull

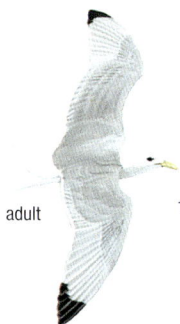
adult

1st-winter

non-breeding

Black-legged Kittiwake

non-breeding

breeding

Ross's Gull

1st-winter

Black Tern *Chlidonias niger*   22–26 cm

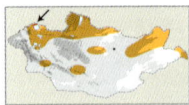

ID Breeding adult confusable only with White-winged Tern, but readily distinguished by dull grey (vs. clear white) upper forewings above and tail, and pale grey (vs. black) underwing-coverts; also, dark (vs. red) legs. Non-breeding adult separated from White-winged by darker mantle; solid black cap and 'sideburns' (vs. pale, streaked cap and cheek spot); grey (vs. whitish) rump; and grey 'halter marks' at sides of breast. Juv distinguished from White-winged by less contrast between back and wings. See also Whiskered Tern. **Voice** At colonies gives a shrill, nasal *kyeeh*; contact call in flight is more abrupt *ki ki* or *klip klip*. **Habitat** Rivers, lakes and large pools with small fish. **Behaviour** Picks food from surface. Light, buoyant flight, usually low over water. **Status** Very rare breeding visitor at Lake Uvs and Sagil and Höndlön River valleys in north-west; and rare passage migrant in suitable habitat south to Valley of the Lakes and east through Hövsgöl and Khangai to Dornod, late April to late August.

White-winged Tern *Chlidonias leucopterus*   20–23 cm

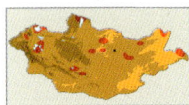

ID Sparkling white wings that contrast sharply with largely black body make breeding adult highly distinctive. For comparison with similar non-breeding adult and juv plumages of Black Tern, see that species (see also Whiskered Tern). **Voice** Gives a soft *kweek* and harsher *kwek kwek* in flight; also a crackling *crzzk*. **Habitat** Nests in marshes and marshy edges of lakes; also frequents these and other aquatic habitats on migration. **Behaviour** Colonial, sometimes nesting with Black-headed Gulls. Feeding behaviour and manner of flight similar to Black Tern. **Status** Locally fairly common breeding visitor from Great Lakes Depression south to Valley of the Lakes and east to Hövsgöl, Khangai and Dornod. Common passage migrant across Mongolia, including lakes and wetlands in arid steppe and Gobi Desert, late April to late August. [Alt· White-winged Black Tern]

Whiskered Tern *Chlidonias hybrida*   23–25 cm

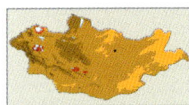

ID Breeding adult resembles a small, darker Common Tern more than other 'marsh terns' (Black and White-winged) with a dark cap, red bill and legs, and grey breast, rather than generally black body plumage. Distinguished from breeding adult Common Tern by smaller size and dark grey underparts, contrasting sharply with white cheeks. Flight pattern more like marsh terns. Non-breeding adult Whiskered also resembles Common Tern, with cap paler than Black Tern but darker and more extensive than White-winged Tern. Juv best distinguished from other marsh terns by brownish-tinged back with heavy black scaling pattern. **Voice** Flight call a dry *kheh* or *grzzt*. **Habitat** Like other marsh terns, nests colonially in reedbeds and tussock sedges along lake and river shores. **Behaviour** Similar to other marsh terns. **Status** Rare and local breeding visitor in Great Lakes Depression and Valley of the Lakes and uncommon passage migrant across Mongolia, including lakes and wetlands in arid steppe and Gobi Desert, late April to mid-August.

Caspian Tern *Hydroprogne caspia*   48–55 cm

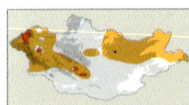

ID The world's largest tern, virtually unmistakable in all plumages due its enormous tomato-red bill. Squarish head shape, short tail and black wing-tips below are also distinctive. **Voice** A harsh throaty *kraaaah* or *kraaaak*. Juv gives whistling *wee-ooo*. **Habitat** Nests colonially on sandy or rocky beaches with sparse vegetation on shores of lakes and large rivers; uses other aquatic habitats on migration. **Behaviour** Plunge-dives for fish; also takes insects and small land animals. **Status** Rare breeding visitor to a number of lakes in Great Lakes Depression and Valley of the Lakes and rare passage migrant in northern and eastern Mongolia, late April to early September. **Taxonomy** Sometimes placed in genus *Sterna*.

breeding

juvenile

Black Tern

non-breeding

breeding

White-winged Tern

juvenile

non-breeding

non-breeding

breeding

breeding

juvenile

Whiskered Tern

non-breeding

breeding

breeding

Caspian Tern

1st-winter

non-breeding

Gull-billed Tern *Gelochelidon nilotica*   35–42 cm

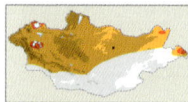

ID Stocky, short-tailed, fairly large tern, best distinguished from other Mongolian terns in all plumages by short, thick, all-black bill. Non-breeding adult has diagnostic black mask/ear patch. Juv has stout dark bill with paler base, and buff tinge on upperparts. **Voice** In flight, gives a brief yapping *kawick*; at colonies a rattling growl. **Habitat** Nests on sandy shores of islands, lakes and rivers; uses other aquatic habitats with other tern species on migration. **Behaviour** Feeds largely on insects, hawking them on wing and picking them from water surface. **Status** Rare and local breeding visitor to lakes Uvs, Khar-Us, Khar and Dörgön (Great Lakes Depression) and Lake Höh in Ulz River valley, Dornod province, and rare passage migrant, mainly to northern Mongolia, south to Valley of the Lakes, late April to to early September.

Common Tern *Sterna hirundo*   34–37 cm

ID Mongolia's most common and widespread tern species. Non-breeding adult, juv and first-winter could be confused with Whiskered Tern, but Common is one-third larger, with notably longer bill, wings and tail. Breeding adult Whiskered Tern has red bill, but in addition to being smaller than Common has dark grey breast and belly, sharply contrasting with white cheeks. Most Mongolian Common Terns belong to black-billed Asian subspecies *S. h. longipennis*, which might be confused with Gull-billed Tern. However, Common is darker grey, has more slender build, much longer tail and lacks Gull-billed's short, heavy bill. Winter adult Common has black crown and nape (vs. Gull-billed's black mask/ear patch. See also Arctic Tern). **Voice** A harsh drawn-out *skreee-arrr*; also a brief *kyik kyik* call. **Habitat** Nests on sandy and rocky shores of rivers and lakes as well as wet meadows, including aquatic habitats in steppe and Gobi Desert. **Behaviour** Colonial breeder sometimes with other gulls and terns. Gregarious on migration, often in mixed species flocks. Buoyant and graceful in flight; hovers and plunge-dives for prey. **Status** Common breeding visitor and passage migrant across Mongolia, late April to early September.

Arctic Tern *Sterna paradisaea*   33–36 cm

ID Very similar to Common Tern, from which it is distinguished by slightly smaller size; rounder head; smaller, all-deep-red bill; shorter legs; and long tail streamers extending well beyond wing-tips at rest (breeding adult). In addition to proportions, juv and winter birds best told by distinct narrow black trailing edge of primaries. **Voice** Shriller and squeakier version of Common Tern's calls. **Habitat** Nests on Arctic shores and winters in southern hemisphere. **Behaviour** Similar to Common Tern. **Status** Vagrant. Two records, a bird at Lake Ögii, Arkhangai province, 5 June 2000, and one at Buur River, Selenge province (date unknown).

Little Tern *Sternula albifrons*   22–24 cm

ID Breeding adult unmistakable because of small size, short tail, yellow bill and white forehead contrasting with black cap. Non-breeding adult and imm distinguished from marsh terns by longer, narrower wings, extensive black in outer primaries, yellowish-brown legs and feeding behaviour. **Voice** A harsh and high-pitched *kek*, *kyik* and *kirik-yik-kirik-yik*. **Habitat** Nests on sandy or pebbly beaches of lake and river shores. **Behaviour** Distinctive jerky, fast-flapping flight. Plunge-dives and often hovers. **Status** Rare and local breeding visitor in Great Lakes Depression, Valley of the Lakes and Lake Buir, Dornod province, and rare passage migrant, late April to late August. **Taxonomy** Formerly placed in genus *Sterna*.

Gull-billed Tern

non-breeding

breeding

non-breeding

juvenile

breeding

Common Tern

breeding

minussensis

longipennis

non-breeding

breeding

breeding

juvenile

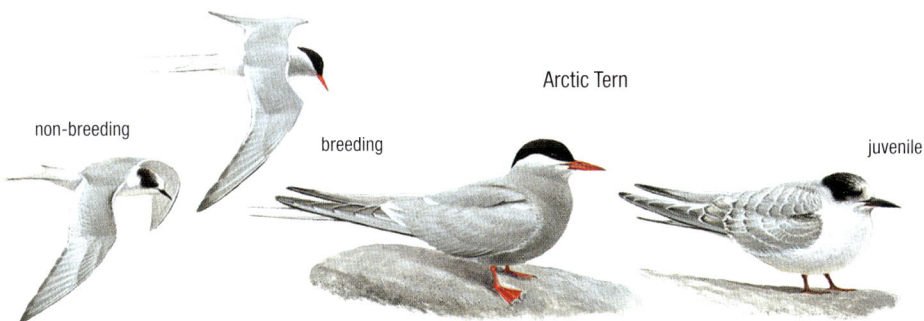

Arctic Tern

non-breeding

breeding

juvenile

Little Tern

non-breeding

non-breeding

breeding

juvenile

breeding

Common Wood Pigeon *Columba palumbus* 38–43 cm

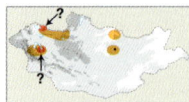

ID Distinctly larger and bulkier than other Mongolian pigeons. At rest, adult shows a white patch on the side of the neck and lacks the black wing-bars of other *Columba* species. In flight, distinguished by prominent white bars at base of flight feathers. **Voice** A monotonous crooning *roo-coo-coo*. **Habitat** Forest and gardens. **Behaviour** Feeds mainly on ground, walking with ponderous waddling gait. Nests in trees. **Status** Rare and very local breeding visitor and rare passage migrant, late April to mid-September at least; some birds overwinter. Nesting is assumed but not confirmed in Tes River valley in Great Lakes Depression; may also nest near Khovd town.

Stock Dove *Columba oenas* 28–32 cm

ID Similar to wild-type Rock and Hill Pigeons but with slighter build, uniform dark grey upperparts, less prominent black bars on wings, and red-and-yellow bill. In flight, wing-tips and trailing edges show more extensive black, and no white rump patch. **Voice** On breeding grounds gives a soft *uhooo* calls. **Habitat** Forest in mountains and river valleys; also parks and gardens. **Behaviour** Feeds on ground. Gregarious when not breeding. Nests in tree cavities. **Status** Rare and very local breeding visitor and rare passage migrant, late April to early September. Nesting confirmed only in Orkhon River valley, Selenge province. Also recorded in Hentii and Khovd provinces.

Yellow-eyed Pigeon *Columba eversmanni* 25–31 cm

ID Similar to other *Columba* pigeons, especially Stock Dove, but paler than that species and with a brownish cast above; yellow eye-ring shows at close range. **Voice** Silent away from breeding grounds. **Habitat** Open country with trees. **Behaviour** Similar to other pigeons. **Status** Vagrant. Two records: one near Bulgan River, Khovd province, 1 May 1975, and one near Lake Telmen, Zavkhan province, 1 June 1998. **Conservation** Vulnerable globally. Listed in Threatened Birds of Asia (2001). [Alt: Pale-backed Pigeon]

Rock Dove *Columba livia* 30–35 cm

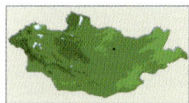

ID Very similar to Hill Pigeon but readily recognised, especially in flight, by absence of white tail-band. Both distinguished from other *Columba* pigeons in Mongolia by pale grey back and wings contrasting with darker grey head and neck. Back colour varies according to race, white in *livia*, pale grey in *neglecta*. Feral Pigeons, flocks of which are common in many cities and towns, exhibit many colour variations: white, dark blue, reddish-brown and mottled. These are descendants of birds bred in captivity and derived from both Rock and Hill Pigeons. **Voice** Courting male gives a three- or four-note *kuku-ru-kooo*; on territory, repeats a drawn-out *kroooo*. **Habitat** Wild birds nest on cliffs and in caves. Feral Pigeons in urban areas nest on buildings, bridges and other artificial structures. **Behaviour** Highly gregarious; nests colonially. **Status** Resident breeder. In Mongolia, wild Rock Doves are found on cliffs in remote areas, notably Uyench River valley in Mongol-Altai range, Bij River in Dzungariin Gobi and Ih Khyangan Mountains in far east. Feral birds occur throughout Mongolia, often in large numbers, especially in cities and towns. [Alt: Rock Pigeon]

Hill Pigeon *Columba rupestris* 33 cm

ID Very similar to Rock Dove, but readily distinguished, especially in flight, by broad white tail-band contrasting with blackish tip. For comparison with other *Columba* pigeons, see Rock Dove. **Voice** Higher-pitched than Rock Dove. **Habitat** Breeds on cliffs, in caves in remote areas and on buildings in small numbers. Feeds in open habitats, including livestock pastures. Feral birds are common in city parks. **Behaviour** Similar to Rock Dove. **Status** Common resident breeder. [Alt: Blue Hill Pigeon]

Common Wood Pigeon

adult

Yellow-eyed Pigeon

Stock Dove

adult

adult

adult

neglecta

livia

adult

Rock Dove

feral pigeons

adult

adult

adult

Hill Pigeon

Eurasian Collared Dove *Streptopelia decaocto*   31–34 cm

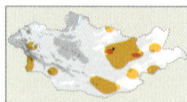

ID Uniform pale buff-grey plumage and black neck bar are diagnostic. **Voice** Male gives a three-note crooning *coo coooo coo*. Flight call a high emphatic *crooooon*. **Habitat** Deciduous forest in mountain and river valleys, stands of trees near agricultural land, and parks and gardens in towns. **Behaviour** Forages on the ground mainly for plant material; often tame and approachable. **Status** Rare and very local resident breeder and passage migrant, expanding its range since 1980s.

European Turtle Dove *Streptopelia turtur*   25–27 cm

ID Distinguished from commoner Oriental Turtle Dove by smaller size and paler underparts coloration and upperwing feathers, which have smaller dark centres and broader rufous fringes. See also Oriental Turtle Dove. **Voice** Song is a soothing purr, *turr, turr*. **Habitat** Nests in deciduous and mixed forest and feeds in brushland and other open areas. **Behaviour** Often roosts in trees, but feeds mainly on ground. Usually shy. **Status** Very rare and local breeding visitor in Bulgan and Bondonch River valleys, Khovd province, where it may overwinter. Also recorded at Segs Tsagaan Bogd Mountain in Trans-Altai Gobi and in mountains north of Lake Hövsgöl. **Conservation** Considered Vulnerable globally. [Alt: Eurasian Turtle Dove]

Oriental Turtle Dove *Streptopelia orientalis*   30–35 cm

ID Larger and stockier than much rarer Eurasian Collared and European Turtle Doves. Scaly feathers of mantle have larger dark centres and narrower rufous fringes than Eurasian, creating an overall darker appearance. Subspecies *S. o. orientalis* has grey, not white, tail-tip; subspecies *S. o. meena* has white tail-tip like European Turtle Dove. Juv similar but duller. **Voice** Song is a repeated crooning *doo-doo doo-doo hoo-hoo*. **Habitat** Nests in deciduous and mixed forest in mountains and river valleys, but occurs in parks, gardens and oases in desert steppe and Gobi on migration. **Behaviour** Spends more time on ground than European Turtle Dove. **Status** Fairly common breeding visitor in northern third of country and fairly common passage migrant throughout, late April to late August (early September in the Gobi). *S. o. meena* occurs in the Great Lakes Depression and Mongol–Altai mountain range in the west, and *S. o. orientalis* in the rest of the country. [Alt: Rufous Turtle Dove]

Laughing Dove *Spilopelia senegalensis*   23–26 cm

ID Smaller, more slender and longer-tailed than *Streptopelia* doves in Mongolia. It is also distinguished by unscaled reddish-brown upperparts, reddish and black speckled 'necklace' and absence of black on undertail. **Voice** Low chuckling *ooo-took-took-ooo-croo*. **Habitat** Deciduous and mixed riparian forest. **Behaviour** Forages mainly on ground. Often tame and approachable. **Status** Very rare and local breeding visitor and presumed winter resident in isolated localities in Bulgan River Valley, Khovd province, and rare passage migrant, late April to early September. **Taxonomy** Formerly placed in genus *Streptopelia*.

adult

juvenile

adult

Eurasian Collared Dove

adult

juvenile

adult

European Turtle Dove

adult

juvenile

meena   adult

Oriental Turtle Dove

adult

Laughing Dove

Lesser Cuckoo *Cuculus poliocephalus*
25 cm

ID Best distinguished from similar Common and Oriental Cuckoos by size (almost one-third smaller) and by finer bill. **Voice** Male's territorial call is an emphatic six-note whistle: *we-choo-tee-we-choo-tee*. **Habitat** Forest. **Behaviour** A brood parasite of *Cettia* warblers. **Status** Vagrant. One in Delgermörön River valley, Hövsgöl province, August 1995.

Indian Cuckoo *Cuculus micropterus*
32–33 cm

ID Distinguished from very similar Common and Oriental Cuckoos by darker grey back; broader more widely spaced barring below; and black terminal band and more extensive white speckling on tail. **Voice** Male's territorial song is a repetitive, mellow, four-syllable phrase with emphasis on third note, often rendered as 'one more bottle'. **Habitat** Well-wooded country. **Behaviour** Elusive. A brood parasite of drongos. **Status** Vagrant. One in Nömrög River valley, Dornod province, 6 June 1995.

Oriental Cuckoo *Cuculus optatus*
30–34 cm

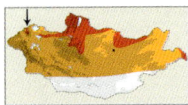

ID Generally not distinguishable by plumage in the field from Common Cuckoo, which see for subtle distinctions. **Voice** Male gives a *pu-pu-pu-pu-pu*...song very similar to that of Eurasian Hoopoe, with which it widely co-occurs, and very unlike Common Cuckoo's *cuoo-ku*. **Habitat** Coniferous, deciduous and mixed forest in Mongol-Altai, Hövsgöl, Khangai and Hentii mountain ranges. Also occurs in arid steppe on migration. **Behaviour** As Common Cuckoo; parasitises *Phylloscopus* warblers. **Status** Uncommon breeding visitor and fairly common passage migrant, late April to late August. **Taxonomy** Formerly treated as *C. saturatus* when united with extralimital Himalayan Cuckoo.

Common Cuckoo *Cuculus canorus*
32–36 cm

ID Virtually identical to Oriental Cuckoo, though the latter typically has heavier black barring above and below in both grey and rufous morphs, and has yellowish (vs. white) undertail-coverts. However, these characters are variable and hard to discern in the field and the two species are best distinguished by song. **Voice** Male constantly repeats familiar cuckoo clock *cuoo-ku* song – very different from Oriental Cuckoo's monotone *pu-pu-pu-pu-pu*.... Female has bubbling call. **Habitat** Coniferous, deciduous and mixed forests and woodlands. Occurs in almost all habitats on migration. **Behaviour** A brood parasite that lays its eggs in nests of small songbirds such as warblers, accentors and pipits, which rear the cuckoo's young at expense of their own. **Status** Fairly common breeding visitor in Hövsgöl, Khangal and Hentii mountain ranges and passage migrant throughout, late April to late August. [Alt: Eurasian Cuckoo]

♂

juvenile

♀
hepatic

Lesser Cuckoo

♂

Indian Cuckoo

juvenile

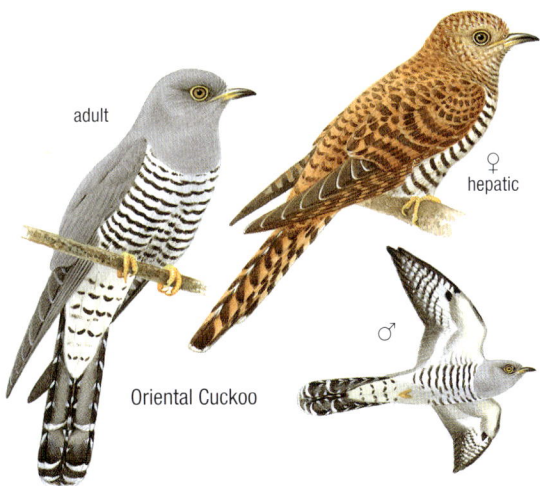

adult

♀
hepatic

♂

Oriental Cuckoo

juvenile

♂

♂

Common Cuckoo

♀
hepatic

juvenile

Snowy Owl *Bubo scandiacus*   53–65 cm

ID A very large white owl with variable amounts of dark spotting in females and imm birds. **Voice** Silent away from breeding grounds. **Habitat** Alpine barrens, mountain steppe, steppe and valleys of large rivers and lakes. **Behaviour** Active by day. Predates small to medium-sized mammals and birds. **Status** Rare winter visitor from Arctic to northern Mongolia, late November to late April. Numbers and movements vary according to food availability. Circumpolar breeding population appears stable. **Taxonomy** Formerly placed in genus *Nyctea*.

Eurasian Eagle Owl *Bubo bubo*   56–75 cm

ID Combination of very large size, prominent ear-tufts and orange eyes distinguish this from other Mongolian owls; compare Long-eared Owl. **Voice** On territory, gives a far-carrying, very deep, hoarse coughing *uh-hoo* or *whoohuu*, repeated infrequently; also a variety of barking calls. **Habitat** Nests on cliffs, sandy precipices, banks of rivers and lakes, abandoned buildings and trees. **Behaviour** Mainly nocturnal. Very powerful predator of birds (including Raven and Saker Falcon as well as smaller species) and mammals. **Status** Common resident breeder throughout Mongolia. [Alt: Northern Eagle Owl]

Long-eared Owl *Asio otus*   31–37 cm

ID Moderate size, slender build and rich tawny facial disc outlined in black distinguish this species from much larger Eurasian Eagle Owl. Similar Short-eared Owl is paler overall with yellowish facial disc and yellow (not orange) eyes and differs markedly in habits (see below). Short-eared Owl rarely shows vestigial ear-tufts compared to long vertical tufts of Long-eared (though latter not always held erect). In flight, Long-eared has streaked belly (mostly white in Short-eared) and lacks the latter's black-barred wing-tips and tail. **Voice** On territory, male gives a groaning *boo boo boo*, but not as vocal as other nocturnal owls; also short, nasal alarm notes. Female gives a higher-pitched soft *shiooosh*. Fledglings have a characteristic 'squeaky-gate' begging call. **Habitat** Forest and patchy woodland in mountains and river valleys, very different from open habitats favoured by Short-eared Owl. **Behaviour** Nocturnal. Shy and elusive. Nests in abandoned corvid or raptor nests. Preys mainly on small mammals. When discovered in day roost, often assumes an exaggerated erect posture next to tree trunk as camouflage. **Status** Uncommon breeding visitor, nesting in forested (mainly northern) parts of Mongolia and passage migrant, mid-April to mid-September. Few individuals overwinter in some areas.

Short-eared Owl *Asio flammeus*   33–40 cm

ID Similar to Long-eared Owl, but paler overall with yellow (vs. orange) eyes, white (vs. streaked) belly and heavy black barring on wing-tips and tail in flight. Behaviour also very different (see above). **Voice** Territorial male gives a soft, low *uh uh uh* hoots; other calls include a more yelping *gi-yow*. **Habitat** Open habitats such as marshes, bogs, shrubby steppe, wheat fields and woodland edges, sometimes roosting in patchy woodlands in mountains and river valleys. **Behaviour** Often hunts during day, quartering open ground like a harrier. Preys mainly on small mammals, also birds. **Status** Uncommon breeding visitor and passage migrant, occasionally overwintering in forested areas and plantations, depending on food availability. Mainly mid-April to mid-September.

Snowy Owl

♂

♀

Eurasian Eagle Owl

adult

adult

Long-eared Owl

adult

Short-eared Owl

Eurasian Scops Owl *Otus scops* 19–21cm

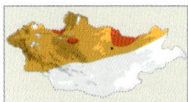

ID Distinguished from other small owls by relatively slender form, prominent ear-tufts and distinctive, complex plumage pattern that blends cryptically with tree bark. Occurs in grey and brown colour morphs. Grey morph is not safely distinguishable in field from grey form of Oriental Scops Owl; brown form of latter is notably more rufous. However, the calls of the two species are very different, and there is little if any overlap in their ranges in Mongolia. Longer primary projection beyond tertials in Eurasian Scops Owl may be visible at close range. **Voice** A plaintive bell-like whistle, *tyuu. . . tyuu. . .* or *duh...duh...duh*, repeated without interruption for many minutes. **Habitat** Forest in mountains and river valleys; also occurs in wooded parks and gardens. **Behaviour** Nocturnal. Feeds mainly on insects and small rodents. Nests in tree cavities. **Status** Common breeding visitor and passage migrant, late April to early September, in forested areas of the north and west, especially Hentii and Khangai mountains.

Oriental Scops Owl *Otus sunia* 19 cm

ID Probably not safely distinguishable from Eurasian Scops Owl by sight in the field, though brown colour form of Oriental is more rufous, and all forms have a shorter primary projection. However, calls and Mongolian ranges are very different. **Voice** Trisyllabic phrase, *bu ku-so*, soft (especially first syllable) and repeated several times; described as frog-like. **Habitat** Forest in river valleys and patchy woodlands in eastern Mongolia. **Behaviour** Similar to Eurasian Scops Owl. **Status** Very rare passage migrant and possible breeding visitor in the east (Khalkh and Nömrög River basins), late April to early September.

Boreal Owl *Aegolius funereus* 22–27 cm

ID Adult distinguished from other small owls in Mongolia by seemingly oversized, squarish head, heavy black border encircling white facial disc and 'eyebrow' marks that create startled expression. Juv dark chocolate-brown all over with some white spotting in wings and on tail, and white facial markings. **Voice** Territorial male gives a rapid series of short, quiet hoots, *popopopopo*. **Habitat** Mature boreal forest. **Behaviour** Nocturnal and shy. Uses old woodpecker holes, especially those made by Black Woodpecker, as nesting cavities. Preys on insects and small mammals. **Status** Uncommon resident breeder in northern taiga and mountain ranges. [Alt: Tengmalm's Owl]

Little Owl *Athene noctua* 23–28 cm

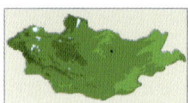

ID Small, plump owl with large rounded head and no ear-tufts. Could be confused with much smaller Eurasian Pygmy Owl but lacks that species' barred flanks. Also differs greatly in behaviour (see below). **Voice** Whistling, repeated *queeou* or *kew*; also gives series of quiet, low-pitched hoots. **Habitat** Favours open habitats in steppe, mountains and desert; often encountered near human settlements. **Behaviour** Typically perches prominently in daylight on rocks, trees, utility poles, outbuildings or the ground. Hunts small rodents, small passerines, lizards and insects in low light around dawn and dusk, when call is most often heard. Nests in rock crevices, tree holes, buildings, etc. **Status** Common resident breeder throughout Mongolia.

Eurasian Pygmy Owl *Glaucidium passerinum* 15–19 cm

ID A tiny, plump owl (size of White-cheeked Starling), 30% smaller than Tengmalm's and Little Owls. Heavily barred throat and flanks (smoky-brown in juv) are also diagnostic. Makes direct, rapid undulating flights like a woodpecker. **Voice** A rapid series of 10–15 sharp breathy whistles *cheuk, cheuk...* that are easily imitated; often calls from prominent perches. **Habitat** Mature coniferous and mixed forest. **Behaviour** Active mainly at dawn and dusk. Nests in woodpecker holes. Feeds on small rodents, songbirds and insects. **Status** Uncommon resident breeder in northern Mongolian forested areas and extreme east of Dornod province.

adult

brown morph

Eurasian Scops Owl

adult

grey morph

Oriental Scops Owl

adult

Boreal Owl

juvenile

adult

Little Owl

adult

Eurasian Pygmy Owl

Northern Hawk-Owl *Surnia ulula* 35–43 cm

ID Distinguishable from all other Mongolian owls by long, pointed tail, giving it an accipiter-like silhouette; uniformly barred (vs. streaked) underparts; and broad, black 'sideburns' framing white face. Juv like adult but duller, with less distinct markings. Makes direct flights with rapid wing-beats like a sparrowhawk. **Voice** Rapid trilling *popopopopopo* on territory; also a falcon-like chittering alarm call. **Habitat** Open conifer and mixed forest in taiga in breeding season and river valleys in winter. **Behaviour** Hunts largely by day; typically chooses prominent lookout perches at tops of trees overlooking open areas. Nests in tree cavities or old raptor nests. Preys on small to medium-sized birds and mammals. **Status** Uncommon resident breeder in north and far east.

Ural Owl *Strix uralensis* 50–59 cm

ID A large, pale, greyish-brown owl, distinguished from all other Mongolian owls by its small black (vs. yellow or orange) eyes. Lacks Great Grey Owl's white facial crescents and broad terminal tail-band. **Voice** Low, throaty hoots, *gu hu hu hoo hoo hoooo*. Female has Grey Heron-like harsh croaky shriek. **Habitat** Mature forest in northern and mountain regions; moves down to river valleys and sometimes enters towns in winter. **Behaviour** Nocturnal but often roosts in open. Preys on rodents and birds. Nests in tree cavities or old raptor nests. **Status** Uncommon resident breeder in forested regions.

Great Grey Owl *Strix nebulosa* 59–68 cm

ID A very large, grey owl with rounded head, long, broad wings, longish tail and notably thick neck. Shape somewhat similar to Ural Owl but much larger with prominent white crescents framing yellow (not dark) eyes. Juv is duller above and below, with indistinct narrow whitish bars on back and underwings. **Voice** Male on territory gives booming hoots, becoming louder *bvoo bvoo bvoo*; female responds with whistling *ehew*. **Habitat** Coniferous and mixed forest in taiga and river valleys. **Behaviour** Hunts in open areas, mainly at dawn and dusk, preying mainly on small mammals. Nests in 'chimneys' of large dead trees or in abandoned raptor nests. **Status** Uncommon and nomadic resident breeder in forested mountain ranges of Hövsgöl and Hentii provinces.

Oriental Dollarbird *Eurystomus orientalis* 30 cm

ID A chunky, iridescent, dark bluish-green bird with red bill and legs, showing pale blue flashes ('silver dollar marks') at base of primaries in flight. Impossible to mistake for any other Mongolian species. Imm has black bill. **Voice** Various harsh calls including *kya kya*, and *shraaak*; cackling *kakakakakaka* in flight. **Habitat** Open country with scattered trees and forest edge. **Behaviour** Perches at top of dead trees and flies out in pursuit of large insects. **Status** Vagrant. One adult in Darkhan district, Hentii province, 17 June 2006, and an adult bird at Khalkh River, Dornod province, 3 June 2014.

adult

adult

Northern Hawk-Owl

adult

Ural Owl

adult

adult

Oriental Dollarbird

Great Grey Owl

Grey Nightjar *Caprimulgus jotaka*   27–29 cm

ID Slightly larger and darker than European Nightjar, but difficult to distinguish either perched or in flight. Best identified by call and distribution. **Voice** Male on territory gives a rapid, accelerating *chut-chut-chut-chut-chut*; also claps wings in flight. **Habitat** Dry rocky slopes, patchy woodlands and open river valleys. **Behaviour** Active from dusk to dawn. Gracefully hawks for moths and other night-flying insects. Rests on ground by day, concealed by cryptic plumage pattern. **Status** Fairly rare breeding visitor and passage migrant to north-central and eastern Mongolia, overlapping with Eurasian Nightjar only in western half of its range, late April to late August. **Taxonomy** Formerly considered a subspecies of extralimital Jungle Nightjar *C. indicus*.

European Nightjar *Caprimulgus europaeus*   24–28 cm

ID Adult paler and slightly smaller than Grey Nightjar, but difficult to distinguish by appearance; best separated by call, narrow widely-spaced dark bars on tail (vs. bold dark bars on Grey Nightjar), and white spot on sides of throat in male (vs. large white throat-patch broken in the centre in male Grey Nightjar) (see above). **Voice** Male on territory gives an incessant purring churr; also soft *kyot* and sharp wing-claps in flight. **Habitat** Dry rocky slopes, patchy woodlands and river valleys. **Behaviour** Similar to Grey Nightjar. **Status** Uncommon to fairly common breeding visitor to northern and western Mongolia, overlapping with Grey Nightjar only in eastern third of its range; uncommon passage migrant throughout Mongolia, late April to late August. [Alt: Eurasian Nightjar]

White-throated Needletail *Hirundapus caudacutus*   20 cm

ID Large size, chunky cigar-shape, blunt-ended tail with small spiky tips, and contrasting white throat and white lower belly distinguish this species from other Mongolian swifts. **Voice** Drawn-out shrill scream. **Habitat** Nests in rock crevices and tree cavities in mountain forests. **Behaviour** An aerial insectivore that rarely perches except when nesting. One of world's fastest fliers, reliably clocked at over 160 km/h in flapping flight. **Status** Very rare and local breeding visitor to northern mountains and lake valleys, and scarce passage migrant in Gobi, late April to late August.

Common Swift *Apus apus*   17 cm

ID Best distinguished from Pacific Swift by absence of that species' white rump. Similarly coloured Eurasian Crag Martin lacks this species' long, deeply forked tail and slender build. **Voice** Noisy, with long, drawn-out shrill screams. **Habitat** Forages over most habitats, including urban areas. **Behaviour** Nests colonially under roofs of buildings, in tree-holes or on cliffs. Often feeds in flocks on aerial insects high overhead. **Status** Locally common breeding visitor and occasionally abundant passage migrant throughout Mongolia, late April to late August. [Alt: Eurasian Swift]

Pacific Swift *Apus pacificus*   15–18 cm

ID Darker than Common Swift, with diagnostic white rump-band and pale scaling on head and underparts. **Voice** Screaming call similar to Common Swift, but somewhat more trilling. **Habitat** Similar to Common Swift. **Behaviour** Similar to Common Swift. **Status** Common to locally abundant breeding visitor and passage migrant throughout the country, late April to late August. [Alt: Fork-tailed Swift]

Grey Nightjar

♂

♀

European Nightjar

♂

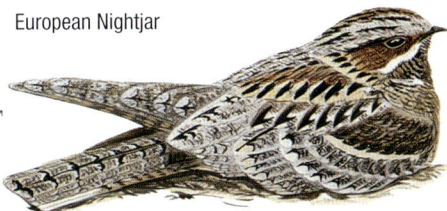

♀

Common Swift

adult

adult

White-throated Needletail

adult

Pacific Swift

Black-capped Kingfisher *Halcyon pileata*   28 cm

ID The only large kingfisher recorded in Mongolia. Combination of black head, blue back, wings and tail, and stout red bill are unmistakable. Shows large white primary-patch in flight. **Voice** A loud cackle *kikikikikikki*. **Habitat** Typically, woodland areas near water. **Behaviour** Feeds mainly on insects but also takes fish and frogs. **Status** Vagrant. A dead bird was found at Baga Gazar Chuluu, Dundgobi province, 15 June 2002. A single individual was photographed near Ongiin Nuuts tourist camp, Ongi River, Övörkhangai province, 30 May 2011.

Common Kingfisher *Alcedo atthis*   16–20 cm

ID A small, colourful kingfisher, unmistakable in Mongolia due to combination of iridescent blue-green plumage above, contrasting with orange breast, belly and ear-patch. Adult male has black bill. Juv duller than adult. **Voice** Piercing whistle given in flight. **Habitat** Shrubby and forested edges of lakes, rivers and pools with small fish. **Behaviour** Plunge-dives for small fish. **Status** Rare breeding visitor in Great Lakes Depression and wetland systems from Selenge province to eastern Dornod, and rare passage migrant in Valley of the Lakes (Lake Bööntsagaan), late April to late August. [Alt: Eurasian/European Kingfisher]

European Bee-eater *Merops apiaster*   28 cm

ID Highly distinctive in both form and spectacular coloration with diagnostic combination of pointed, decurved bill; long, pointed wings and tail; yellow throat; blue underparts; and chestnut crown, mantle and inner wings. Juv duller than adult. **Voice** Call is a rolling *prrrrrp*. **Habitat** Dry, open country. **Behaviour** Hawks insects in graceful flight. **Status** Vagrant. One at Yarantai on the Bulgan River, Khovd province (undated), and two at Ulaangom town, Uvs province, May 2010 and 28 August 2014.

Eurasian Hoopoe *Upupa epops*   25–32 cm

ID Unmistakable with its buff-orange plumage; broad wings, boldly banded in black and white; black-tipped fan-like crest; and long, decurved bill. **Voice** A repetitive and mellow *poop, poop, poop*, similar to song of Oriental Cuckoo. **Habitat** Open habitats of all kinds. **Behaviour** Nests in tree cavities, rock crevices and artificial structures. Forages mainly on ground, feeding on large insects. **Status** Common breeding visitor and fairly common passage migrant throughout Mongolia, late April to late August (early September in the south). [Alt: Common Hoopoe]

adult

Black-capped Kingfisher

♂

♀

Common Kingfisher

European Bee-eater

adult

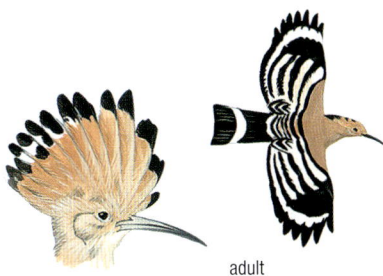

adult

juvenile

Eurasian Hoopoe

Eurasian Wryneck *Jynx torquilla*

16–18 cm

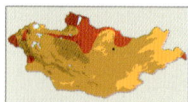

ID A highly atypical woodpecker with colour pattern of a nightjar. Difficult to mistake for any other species once seen, though might at first be taken for a songbird. **Voice** A distinctive, high-pitched, nasal *kwi kwi kwi*, repeated at intervals – often the best way to detect its presence. **Habitat** Nests in coniferous, deciduous and mixed forest, preferring open woodland, riparian thickets, forest edges and isolated copses. Frequents open habitats with scattered (incl. planted) trees, and rocky hillsides with trees and tall bushes on migration. **Behaviour** Forages for ants and ant pupae in trees and on ground, but does not inch up tree trunks using tail as a prop (cf. typical woodpeckers). In flight, flaps and glides like a shrike rather than undulating like other woodpeckers. Does not drum, nor excavate its own nest-holes but uses cavities made by other woodpeckers. Often sits very still, with its neck 'awry', appearing like part of the tree. **Status** Fairly common breeding visitor in forested north and mountainous regions of country and uncommon passage migrant throughout, late April to late August (early September in the Gobi).

Grey-headed Woodpecker *Picus canus*

27–30 cm

ID Coloration unlike any other Mongolian woodpecker, with grey head, olive-green mantle and tail, and brighter brass-yellow rump. Male has red forehead patch that female lacks; young birds duller than adults. **Voice** A soft, plaintive whistle, *pio-pio-pio-pio-pio*, up to 20 notes; also, single *kik*. Drumming rather loud and fast, and in 'rolls' of up to 1.5 seconds. **Habitat** Coniferous and mixed forested areas, descending to forest steppe and river valleys in winter when it sometimes visits parks and gardens. **Behaviour** Feeds mainly on tree trunks and higher branches; also sometimes on the ground. **Status** Uncommon resident breeder in forested regions. [Alt: Grey-faced Woodpecker]

Black Woodpecker *Dryocopus martius*

40–46 cm

ID A crow-sized black bird, far larger than any other Mongolian woodpecker and with no white in its plumage. Females and juvs have less red on crown than adult male. **Voice** Usual call is a ringing *kyoon kyoon*, also a shrill and loud *kreeo kreeo kreeo* and a rapid *kiokiokiokio* in flight. Drumming very loud and forceful ('like machine gun') up to 3 seconds long. **Habitat** Coniferous and mixed forest in taiga, descending to forest steppe and river valleys in winter. **Behaviour** Shy and easily overlooked despite its size. Feeds mainly on rotting tree trunks, making large distinctive excavations in search of carpenter ants and wood-boring beetle larvae. Arguably a 'keystone' species as its holes are used by many other birds and mammals. **Status** Fairly common resident breeder in forested areas.

Eurasian Three-toed Woodpecker *Picoides tridactylus*

22 cm

ID Adult male distinguished from other Mongolian woodpeckers by lemon-yellow crown patch. Female and juv distinguished by black-streaked crown, prominent white panel from nape to rump, and black face with white stripes (rather than reverse in other species). **Voice** An abrupt *kick* contact call, not as hard as Great Spotted Woodpecker's. Alarm call a short, rattling trill: *kli-kli-kli....* Drumming bursts short, but strong, accelerating slightly towards end. **Habitat** Coniferous and mixed forest in taiga, descending to forest steppe and river valleys in winter. **Behaviour** Drills small holes for sap as well as flaking off wood to search for wood-boring grubs. Forages mainly on decaying conifers. **Status** Uncommon resident breeder, mainly in north of country.

adult

Eurasian Wryneck

♀

♂

♀

Grey-headed Woodpecker

♂

Black Woodpecker

♂

♀

Eurasian Three-toed Woodpecker

Great Spotted Woodpecker *Dendrocopos major*   23–26 cm

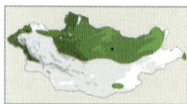

ID Distinguished in all plumages from similar White-backed and much smaller Lesser Spotted Woodpeckers by pair of elongated white shoulder patches, conspicuous both at rest and in flight; also lacks White-backed's white rump. **Voice** Flight call is a hard emphatic *kick*, sometimes repeated rapidly, *kikikikik*. Drumming brief (*c.* 1 second) and very rapid, sounding like creaking branch. **Habitat** Coniferous, deciduous and mixed forest in taiga, descending to forest steppe, and river valleys in winter when often appears in urban parks and gardens or (rarely) around cliffs in steppe. **Behaviour** Probes tree trunks for insects and feeds heavily on conifer seeds in winter. **Status** Common resident breeder, mainly in north of country, and rare winter visitor from farther north.

White-backed Woodpecker *Dendrocopos leucotos*   25–30 cm

ID White rump patch is diagnostic. Barring on upperparts recalls Lesser Spotted, but White-backed is much larger; also lacks the large pair of white shoulder patches of similar Great Spotted Woodpecker. **Voice** Main calls are a soft *wick* and a harder *kyo kyo*. Drumming loud and accelerating, fading away at end; bursts last 2 seconds. **Habitat** Coniferous, deciduous and mixed forest in taiga, often near water; moves down to forest steppe, river valleys and city parks and gardens in winter. **Behaviour** Prefers mature, undisturbed forest, typically feeding low on decaying trees and making large craters while probing for insect larvae. **Status** Fairly common resident breeder in forested regions, though not as conspicuous as Great Spotted. *D. l. uralensis* occurs in NW Mongolia, nominate race elsewhere.

Rufous-bellied Woodpecker *Dendrocopos hyperythrus*   20–25 cm

ID Readily distinguished in all plumages from regularly occurring Mongolian woodpeckers by bright rufous coloration on head, breast and belly. **Voice** Call is a rapid *tik-tik-tik* or *tiktiktiktik*. Drumming brief, accelerating. **Habitat** Deciduous forest and woodland. **Behaviour** Similar to Great Spotted Woodpecker. **Status** Vagrant. Two records: a single individual at Ikh Nart Nature Reserve, Dornogobi province, 3 September 2005, and an adult in Tuul River valley, Uubulan, Töv province, 18 September 2007.

Lesser Spotted Woodpecker *Dryobates minor*   14–16 cm

ID Mongolia's smallest woodpecker (little more than half size of Great Spotted) with tiny bill; black-and-white barred back also distinctive. Adult male has red crown, lacking in female. **Voice** Call like Great Spotted but weaker; 'song' a sharp series, *kee-kee-kee...*. Drumming weak, often making two 'rolls' in quick succession. **Habitat** Coniferous, deciduous and mixed forest in taiga, descending to forest steppe, river valleys and gardens in winter. **Behaviour** Agile, feeding mainly on small outer twigs and branches. Occasionally visits reedbeds. **Status** Fairly common resident breeder across forested areas of northern Mongolia, and rare winter visitor from further north. **Taxonomy** Formerly placed in genus *Dendrocopos*.

Great Spotted Woodpecker

♀

♂

uralensis

♀

♂

leucotos

♀

♂

White-backed Woodpecker

♂

♀

Rufous-bellied Woodpecker

♂

♀

Lesser Spotted Woodpecker

Bull-headed Shrike *Lanius bucephalus*
20 cm

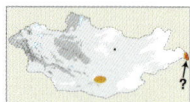

ID Best distinguished from similar Brown and Isabelline Shrikes by combination of rufous cap and grey back and tail in adult male, and scaly barring below and indistinct mask in adult female and 1st-winter. Bull-headed is chunkier with comparatively large head. See also status, below. **Voice** Usual call is a chattering *chu-chu-chu*. Song is a raspy *chiki-chiki-chiki*, ending with longer notes and sometimes including some mimicry. **Habitat** Shrubby areas in steppe and desert. **Behaviour** Pounces on prey from open perch. Feeds mainly on small invertebrates and lizards. **Status** Very rare passage migrant in eastern Mongolia. Also recorded in Ömnögobi, and may breed in Nömrög River valley, Dornod province.

Brown Shrike *Lanius cristatus*
20 cm

ID Most likely to be confused with much paler Isabelline Shrike, but darker overall and lacks that species' white primary patch and contrasting bright rufous tail. Compare also much rarer Bull-headed Shrike (see distribution under Status, below). **Voice** A harsh, rapid *che-che-che*. **Habitat** Bushes and small trees on forest margins in mountain steppe and river valleys. **Behaviour** Like other shrikes, pounces on prey from prominent perch. Feeds mainly on insects, also lizards and small mammals; may store prey by impaling it on thorns. **Status** Common breeding visitor and passage migrant throughout Mongolia, late April to mid-August (early September in the Gobi).

Isabelline Shrike *Lanius isabellinus*
16–18 cm

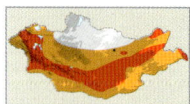

ID Readily distinguished from other small shrikes in Mongolia by pale sandy grey-brown upperparts, contrasting rufous tail and prominent white primary patch. Adult female and 1st-winter/juv birds have fine scaling on breast and/or flanks. **Voice** A dry, harsh *chack chack*. **Habitat** Dry, open habitats with scattered tall shrubs and small trees; also dense willows and other tall bushes in dry river valleys, and planted poplar and other deciduous trees in towns and cities. Partially replaces Brown Shrike further south in arid steppe and desert. **Behaviour** Similar to Brown Shrike. **Status** Fairly common breeding visitor and passage migrant in dry, open habitats across Mongolia, late April to late August (early September in the south). [Alt: Rufous-tailed Shrike]

Long-tailed Shrike *Lanius schach*
21–25 cm

ID Generally resembles larger grey shrikes but distinguished by rufous flanks, undertail-coverts, lower mantle and rump. **Voice** Commonly a harsh, grating *shaak, shaak*. **Habitat** Open country with shrubs and small trees. **Behaviour** Similar to larger shrikes. **Status** Vagrant. Two in Gurvansaikhan mountain range, Ömnögobi province, 27 May 1986.

♂ ♀

juvenile

Bull-headed Shrike

♀ ♂

♂

1st-winter

Isabelline Shrike

1st-winter

Brown Shrike

♂

juvenile

Long-tailed Shrike

Lesser Grey Shrike *Lanius minor*   19–21 cm

ID Distinguished from larger grey shrikes by smaller size, more compact build and especially by extension of black face-mask above the bill and onto the forecrown. Adult male also has pink wash on lower breast and belly. Female like male but duller. **Voice** A babbling song that may include mimicry of birds and even mammals; also typical shrike *tchak, tchak* alarm calls. **Habitat** Open areas with scattered tall shrubs and small trees. **Behaviour** Typical shrike hunting style. Prefers large insects and small rodents. **Status** Very rare and local breeding visitor; known from single nest site in Bulgan, Khovd province, 24 June 2006.

Northern Shrike *Lanius borealis*   22–26 cm

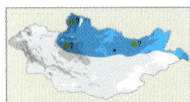

ID This and the following two species (Steppe and Chinese Grey Shrikes) differ from all other Mongolian shrikes by their large size, clean grey, black and white plumage pattern and absence of any rufous or pinkish tones. Adult is slightly darker above than Steppe Grey Shrike, with a small white patch at base of black primaries. Underparts are tinged buffish (*L. b. mollis*) or pure white (*L. b. sibiricus*), but usually finely barred. Juv/1st-winter is brownish-grey above with fine brownish barring below and pale bill. **Voice** Typically silent when not breeding, though gives occasional harsh nasal bleating calls. **Habitat** Nests in small trees or tall shrubs at forest edges in taiga forest; winters in open steppe with scattered trees and bushes. **Behaviour** Maintains lookouts from prominent perches from which it makes sallies after small mammals and other prey; often hovers. **Status** Possible resident breeder in Khangai, Hentii and Hövsgöl mountain ranges and regular winter visitor down to forest steppe. **Taxonomy** The larger grey shrikes have been the subject of much taxonomic confusion with numerous subspecies recognised. Accordingly, this and the following species have sometimes been lumped as subspecies of Great Grey Shrike *L. excubitor*.

Steppe Grey Shrike *Lanius pallidirostris*   24–25 cm

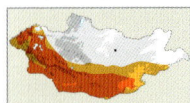

ID Adult has narrow black lores (vs. wider in Northern and extralimital Great Grey), large white patch at base of closed primaries (vs. smaller in Northern), and conspicuous white tips to secondaries. Bill is black in male and pale with a dark tip in female. Juv/1st-winter is paler than adult with reduced mask and mainly pale lores. Underparts are unbarred (vs. finely barred in Northern) and bill is very pale with dark grey tip. **Voice** Song is highly variable and rather quiet with combinations of trills, chatters and squeaky notes; calls as Northern Shrike. **Habitat** Nests and forages in desert steppe and Gobi desert with scattered Saxaul trees and tall *Caragana* bushes. **Behaviour** As Northern Shrike. **Status** Fairly common breeding visitor in suitable habitat from the NW Altai and Great Lakes Depression south and east through the Gobi desert, and uncommon passage migrant throughout the southern half of the country; mid-April to mid-September. **Taxonomy** See under Northern Shrike, above.

Chinese Grey Shrike *Lanius sphenocercus*   31 cm

ID Generally similar to the preceding two species, but larger with longer tail and more extensive white wing patches. **Voice** As Steppe Grey Shrike. **Habitat** Bushy areas in arid steppe, desert and river valleys with sandy soil, including dunes, with small poplars or other deciduous trees. **Behaviour** As Northern Shrike. **Status** Rare and very local breeding visitor and rare passage migrant with a single confirmed nesting record from Khalkh River valley, Dornod province, June 2013; also single birds photographed in Saxaul trees in Nomgon district, Ömnögobi province, June 2005; in Khustai National Park, Töv province 7 and 24 September 2010; and in Khalzan district, Sukhbaatar province, August 2011.

juvenile

♀

♂

Lesser Grey Shrike

mollis

adult

Northern Shrike

sibiricus

1st-winter

adult

pallidirostris

1st-winter

♀

Steppe Grey Shrike

♂

pallidirostris

adult

Chinese Grey Shrike

Black Drongo *Dicrurus macrocercus*

27–30 cm

ID The only all-black Mongolian bird with long, slightly upwards-curved, deeply-forked tail and small white spot on the gape. 1st-winter has less forked and uncurved tail and some whitish scaling above and below. **Voice** Rasping monosyllables, *sheea, tzeee*. **Habitat** Open country with trees. **Behaviour** Flycatches from prominent perches in agile sallies. **Status** Vagrant. One at Juulchin Gobi tourist camp, Ömnögobi province, 22 June 2000 and eight birds in a poplar plantation in Dalanzadgad, Ömnögobi province.

Eurasian Golden Oriole *Oriolus oriolus*

22–25 cm

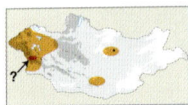

ID In all plumages unlike any other Mongolian bird apart from vagrant Black-naped Oriole. Bright gold or greenish plumage and dark wings with heavy red bill distinguishes all but dull juv, in which stout pinkish bill, fine breast streaking, and olive-green coloration taken together are definitive. Compare Black-naped Oriole. **Voice** Song is loud, melodious and fluty; calls harsh and shrill. **Habitat** Deciduous and mixed forest in mountains and river valleys. **Behaviour** Shy. Nests and forages discreetly in tree-tops and feeds mainly on insects; also berries. **Status** Very rare breeding visitor and passage migrant mainly in western Mongolia, with nesting recorded in Bulgan River valley, Khovd province, and migration observed through Great Lakes Depression; also a single bird in Hustai Nuruu National Park, Töv province, 21 July 2007. Adult male and female in planted deciduous trees at Dungenee Brigad, a small settlement in the Ömnögobi province, 5 July 2015. Presumably present late April to late August.

Black-naped Oriole *Oriolus chinensis*

26–27 cm

ID Adults of both sexes readily told from Eurasian Golden Oriole by black 'bandana' from lores to nape. Juv is distinctly yellow (not olive) with little or no black in wings, heavier breast streaking, stouter bill, and at least a trace of bandana. **Voice** Similar to Eurasian Golden Oriole but a little lower-pitched. **Habitat** Deciduous forest in river valleys. **Behaviour** Similar to Eurasian Golden Oriole. **Status** Vagrant. Single individuals at Lake Buir and Nömrög River, Dornod province, 8 June 1995.

juvenile

Black Drongo

adult

♀

♂

Eurasian Golden Oriole

♂

♀

juvenile

Black-naped Oriole

PLATE 63: STARLINGS & MYNA

Daurian Starling *Agropsar sturninus*   18 cm

ID Adult male's purple-and-green back and wings, contrasting with pale grey head and underparts, are diagnostic. Adult female and juv have brown back and wings, but small size, grey plumage, beady black eye and black legs and feet are distinctive. Compare juv Rosy and Common Starlings. **Voice** Song is a typical starling medley of squeaks, rattles and mimicry of other species; call a throaty *kryoo*. **Habitat** Nests in tree cavities in deciduous woodland in forest steppe and river valleys. **Behaviour** Feeds on fruit, seeds and invertebrates in trees and on ground. Forms post-breeding migratory flocks. **Status** Uncommon and local breeding visitor in Khalkh and Nömrög River basins in Dornod province and scarce passage migrant in Hentii and Khangai ranges, late April to early September. **Taxonomy** Formerly placed in genus *Sturnia*. [Alt: Purple-backed Starling]

White-cheeked Starling *Spodiopsar cineraceus*   24 cm

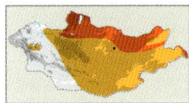

ID In all plumages, the prominent white cheek-patch, combined with white rump and yellowish bill and legs, is diagnostic. Juv is duller than adult. **Voice** Unmusical song rendered as *chir-chir-chay-cheet-cheet*; call is a harsh *char*. **Habitat** Nests mainly in tree cavities and in deciduous and mixed forests in mountains and river valleys, as well as artificial structures. Forages on open steppe, agricultural fields and desert on migration. **Behaviour** A colonial nester. Very social and gregarious. Feeds mainly on ground, taking insects and seeds, including from droppings of livestock herds. **Status** Fairly common breeding visitor in north and east and fairly common passage migrant over much of rest of country, late April to early September (mid-September in the Gobi). **Taxonomy** Formerly placed in genus *Sturnus*.

Common Starling *Sturnus vulgaris*   19–22 cm

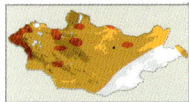

ID Glossy purple-and-green plumage in combination with yellow bill and bright reddish-brown legs is diagnostic in adults. Pale tips to feathers in winter plumage give a distinctive speckled look. Juv resembles juv Rosy Starling, but the latter has thicker, yellowish bill and darker wings, contrasting with paler back and underparts. **Voice** Song is a rather quiet, prolonged and very varied, with rattles, squeaks, whistles and much mimicry of other birds and (in urban areas) mechanical noises. Calls are harsh, guttural *churrr* and *skeee* notes. **Habitat** Woodlands and gardens in forested areas and river valleys, often in proximity to human habitation (including cities). Occurs in arid steppe and desert on migration. **Behaviour** Sociable, lively and quarrelsome. Nests in cavities in trees, rocks and artificial structures. Forages mainly on ground for grubs and other invertebrates, but also takes fruit. **Status** Uncommon to locally common breeding visitor, mainly in montane areas of west and north, and uncommon passage migrant over rest of country, late April to early September. [Alt: European Starling]

Rosy Starling *Pastor roseus*   19–22 cm

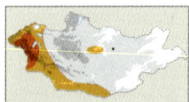

ID Pink 'vest' and crested head render adult unmistakable, even in winter when plumage is duller. Juv resembles that of Common Starling, but note pale bill, contrasting dark wings, less breast streaking and pale rump in flight. **Voice** Song combines mellow warbling phrases with harsher chattering. Most calls are harsh and rasping. **Habitat** Prefers dry, often rocky, areas and steppes with scattered shrubs; also agricultural fields. **Behaviour** Nests in large colonies, typically in rock cavities, stone walls and ruins. Forages mainly on ground for preferred insect prey, but also takes fruits and seeds. Opportunistically follows grasshopper 'plagues', acting as natural pest control. **Status** Uncommon and erratic breeding visitor in western Mongolia, but occurs irregularly as rare migrant and wanderer in the Uvs and Great Lakes Depression and Gobi Desert, late April to early September. **Taxonomy** Formerly placed in genus *Sturnus*. [Alt: Rose-coloured Starling]

Crested Myna *Acridotheres cristatellus*   25–27 cm

ID A large starling. Combination of overall black plumage, frontal crest, yellow-orange eye, ivory bill, yellow legs and white wing-patch and undertail-coverts is unique in Mongolia. Juv is dark brown and lacks crest, but readily recognisable by other characters noted. **Voice** Grating *kruu kruu* notes. **Habitat** Open habitats. **Behaviour** Feeds on ground. May flock with other species. **Status** Vagrant. One found near Khalkh River, Dornod province, June 2000.

♀

♂

Daurian Starling

adult

breeding

White-cheeked Starling

juvenile

non-breeding

Common Starling

breeding

non-breeding

juvenile

Rosy Starling

non-breeding

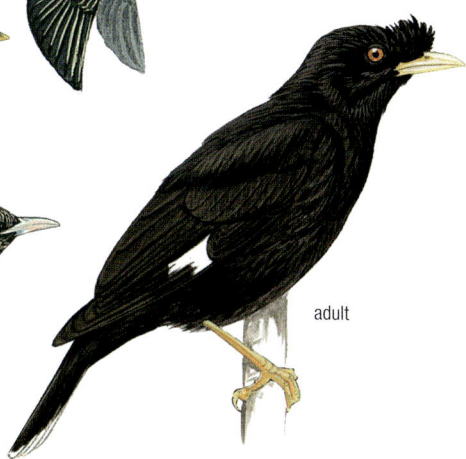

adult

Crested Myna

Siberian Jay *Perisoreus infaustus*

26–30 cm

ID In form, resembles Eurasian Jay with which it shares its range in Mongolia. However, it is notably smaller, darker and plainer, but with distinctive rusty-red outertail feathers, rump and wing patches. **Voice** Generally quiet, sometimes giving soft whistles and buzzard-like mews. **Habitat** Coniferous forest (taiga). **Behaviour** Usually shy and discreet. Somewhat gregarious when not nesting. Omnivorous, taking plant and animal food, including birds' eggs and carrion. Caches food for winter. **Status** Uncommon resident breeder in taiga forest of Hövsgöl, Khangai and Hentii; descends to forest steppe and river valleys in winter. *P. i. opicus* occurs in the west; greyer *P. i. sibericus* is more frequent elsewhere.

Eurasian Jay *Garrulus glandarius*

32–35 cm

ID A striking bird, unlikely to be confused with any other species due to its unusual combination of overall pinkish-buff coloration, highlighted in flight by white and iridescent blue wing-patches, prominent white rump and contrasting all-black tail. **Voice** Very harsh loud scream given in alarm: *skraaak*; also mimics raptors' mewing and cackling calls. **Habitat** Coniferous and mixed forest in taiga and river valleys as well as human settlements. **Behaviour** Rather shy and not very gregarious. Forages on ground and in trees for virtually anything edible; caches pine nuts and similar during winter. **Status** Fairly common resident breeder in Hövsgöl, Khangai and Hentii, overlapping with Siberian Jay but ranging lower into mixed forest.

Azure-winged Magpie *Cyanopica cyanus*

31–35 cm

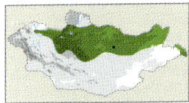

ID Unmistakable. Slender form, pale blue wings, very long tail and contrasting black cap. Juv slightly duller than adult, with brown tinge on cap and breast. **Voice** Various harsh screeches, trills, rattles and whines. **Habitat** Deciduous and mixed woodland and riverine thickets in moister valleys. **Behaviour** Gregarious, usually moving around in small groups, but generally shy and wary. **Status** Uncommon to fairly common resident breeder in Khangai and Hentii east to Dornod.

Eurasian Magpie *Pica pica*

40–51 cm

ID Large size, black-and-white plumage with metallic blue-green highlights in wings and tail, and long tail make this magpie one of Mongolia's most easily recognised birds. **Voice** Harsh rattling *kakakakak* and similar calls. **Habitat** Open wooded habitats (avoids dense forest), thickets, agricultural land with scattered trees, suburban and urban parks and gardens. **Behaviour** Forages on ground with strutting gait; also feeds in trees, taking all kinds of food, including birds' eggs and nestlings. **Status** Common to very common resident breeder across northern half of Mongolia, becoming scarce in treeless, arid south, though nesting locally in Ömnögobi province. [Alt: Common/Black-billed Magpie]

opicus

adult

sibericus

adult

adult

brandtii

Siberian Jay

Eurasian Jay

adult

juvenile

Azure-winged Magpie

adult

Eurasian Magpie

Henderson's Ground Jay *Podoces hendersoni*   30 cm

ID Readily recognised by combination of sandy-buff coloration, glossy black cap, wings and tail, white wing flashes in flight and foraging behaviour. **Voice** Hard, wooden *clack clack clack*. **Habitat** Nests in small Saxaul trees and tall shrubs on rocky hills and mountain valleys in arid steppe and desert. **Behaviour** Runs very rapidly and probes with bill for insect larvae in sandy soil. May perch conspicuously on top of hillocks, rocks and tall bushes, but also adept at concealing itself behind bushes or rocks when approached. In winter, visits winter camps of local herder families near breeding sites. **Status** An iconic bird of Gobi Desert. Locally fairly common to scarce resident breeder in more arid regions of west and south. **Conservation** Considered Vulnerable in Mongolia due to habitat loss. Listed as Threatened in the Mongolian Red Book (2013). [Alt: Mongolian Ground Jay]

Spotted Nutcracker *Nucifraga caryocatactes*   32–35 cm

ID A small arboreal crow with highly distinctive white speckling on chocolate-brown plumage and conspicuous white undertail-coverts and (in flight) white tail-tip. **Voice** Call is deeper, growling version of Eurasian Jay's scream (*skraaak*), sometimes repeated. **Habitat** Coniferous forest in mountains and river valleys. **Behaviour** Quite shy and wary, though often perches on the tops of conifers. Feeds on pine seeds and caches supplies for winter. **Status** Fairly common resident breeder in dense conifer forest from Altai Mountains through Khangai, Hövsgöl and Hentii provinces to Khalkh River-Nömrög region in east. [Alt: Eurasian Nutcracker]

Red-billed Chough *Pyrrhocorax pyrrhocorax*   36–40 cm

ID This glossy black crow with bright red decurved bill and red legs is unmistakable. **Voice** Usual call is a harsh *ch-yow*. **Habitat** Open mountainous areas with grassy openings; also open steppe and urban areas, where it nests in Buddhist temples and other artificial structures. **Behaviour** Feeds on ground, probing with its bill for insect larvae. Gregarious throughout year, often in flocks of several hundred. Very agile in flight, performing communal aerobatic displays. **Status** Common or very common resident breeder in suitable habitat throughout Mongolia.

Western Jackdaw *Coloeus monedula*   30–34 cm

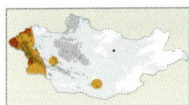

ID A small, compact crow with a short bill, dark grey rather than jet-black plumage, a silvery-grey nape and whitish iris. Only likely to be confused with imm Daurian Jackdaw, which see for distinctions. **Voice** Flocks are noisy. Main call is a high, sharp yelping *jack*; also a harsher *craw* and many other sounds. **Habitat** A cavity nester in tree-holes or buildings. **Behaviour** Omnivorous, feeding mainly on ground and often in flocks, especially after breeding. Agile and aerobatic. **Status** Very rare breeding visitor in Mongol-Altai Mountains and W Great Lakes Depression; very rare passage migrant with records from Ömnögobi province; and rare winter visitor. **Taxonomy** Formerly placed in genus *Corvus*. [Alt: Eurasian Jackdaw]

Daurian Jackdaw *Coloeus dauuricus*   33 cm

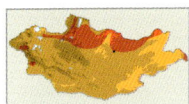

ID The striking black-and-white adult plumage pattern of this small crow is unique in Mongolia. Duller imm and dark morph adult may be confused with Western Jackdaw but Daurian always has black (not whitish) eyes. Adult dark morph birds also have grey patch on side of neck. Compare also Hooded Crow. **Voice** Similar to Western Jackdaw. **Habitat** Cavity nester in mixed and coniferous forest on mountains and river valleys, always in proximity to open steppe or farmland foraging areas. Often found in and near human settlements. **Behaviour** Generally similar to Western Jackdaw. Often flocks with Rooks and Carrion Crows. **Status** Common to fairly common breeding visitor and passage migrant throughout Mongolia, mid-March to late September. Winters in small numbers near breeding sites. **Taxonomy** Formerly placed in genus *Corvus*.

adult

adult

Henderson's Ground Jay

adult

Spotted Nutcracker

adult

adult

Red-billed Chough

Western Jackdaw

adults

adult

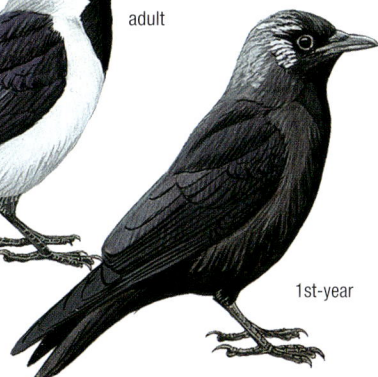

1st-year

Daurian Jackdaw

Rook *Corvus frugilegus*   41–49 cm

ID Similar to Carrion Crow in size and general coloration, but adult distinguished by pale bare skin around bill base, head with a peaked crown and plumage appearing glossier in good light. Juv lacks pale bill base and therefore very like Carrion Crow, but differs in more pointed and straighter bill, suggestion of peaked crown, and call. **Voice** A dry cawing *kraaa*, harsher than call of Carrion Crow. **Habitat** Nests in coniferous, deciduous and mixed forests and parks adjacent to open feeding habitats. **Behaviour** Nests colonially in tree-top 'rookeries'. Gregarious with other crow species when foraging, often in open natural and agricultural fields. **Status** Rather uncommon and local resident breeder and passage migrant throughout Mongolia, nesting mainly in highlands and other forested regions, late April to mid-September; occasionally winters in larger Mongolian cities. *C. f. pastinator*, in most of Mongolia except the west, has less extensive bare skin around the bill base. [Alt: Eurasian Rook]

Carrion Crow *Corvus corone*   44–51 cm

ID Most likely to be confused with two other all-black crows in Mongolia. Northern Raven is significantly larger with a much heavier bill, wedge-shaped tail and distinctive flight profile and call (see below). For distinctions from imm Rook, see that species. **Voice** Similar to Rook but stronger and less harsh; caws repeatedly. **Habitat** Nests in all forest types in proximity to open foraging habitats; will also nest on rocky ledges, on ground (rarely) and in city parks. **Behaviour** Less gregarious than Rook, not breeding in colonies. Forages mainly on ground. Like other corvids, omnivorous and opportunistic. **Status** Common resident breeder across Mongolia, especially where trees and adjacent open habitats provide ideal nesting and feeding conditions.

Hooded Crow *Corvus cornix*   44–51 cm

ID Similar to Carrion Crow except for distinctive grey plumage of nape, back and belly contrasting with black of head, wings, tail and 'bib'. General pattern similar to much smaller Daurian Jackdaw, in which adult is white where Hooded Crow is 'dirty' grey. **Voice** As Carrion Crow. **Habitat** As Carrion Crow. **Behaviour** As Carrion Crow. **Status** Vagrant. Recorded in Hentii and Mongol-Altai mountain ranges. Singles photographed in Khovd and Buyant River valleys and Khovd town, Khovd province; and Bogd Uul and Ulaanbaatar, winters of 2013–16. **Taxonomy** Considered a colour morph of Carrion Crow by many authorities, including BirdLife International; hybrids of intermediate appearance occur.

Northern Raven *Corvus corax*   54–67 cm

ID The largest of the three all-black crows in Mongolia. In addition to overall size this species differs from Carrion Crow and Rook in notably heavier bill; distinctive flight profile with proportionally longer, more pointed wings; a wedge-shaped (not rounded) tail; a more pronounced head projection; and call. **Voice** Distinctive very deep, resonant croak, gives other higher-pitched calls from time to time. **Habitat** Nests on cliffs, in trees, and on utility poles, city buildings and other artificial structures. **Behaviour** Agile and apparently playful on the wing, pairs often performing dramatic tumbles and rolls. Not as gregarious as other crows. Omnivorous. **Status** Very common resident breeder throughout Mongolia. [Alt: Common Raven]

adult

Rook

adult

frugilegus

pastinator

juvenile

Carrion Crow

adult

adult
calling

adult

Hooded Crow

adult

Northern Raven

Bohemian Waxwing *Bombycilla garrulus* — 18–21 cm

ID Form and plumage pattern – especially prominent crest and broad bright yellow terminal tail-band – make this species impossible to confuse with any other Mongolian bird except the vagrant Japanese Waxwing, which see for distinctions. **Voice** Winter flocks very vocal; call is a sibilant high trilling *sriiiiiii sriiiiiiiii*. Slow-paced song combines similar trills with more grating sounds. **Habitat** Nests in coniferous and mixed forest, often near water. Wintering birds seek out berry-bearing trees both in rural areas and in towns. **Behaviour** Feeds on insects in breeding season, switching to fruit in autumn and winter. Irruptive, gregarious and semi-nomadic, with flocks from further north arriving in varying numbers to winter. **Status** Presumed very rare and local breeding visitor to Khangai, Hövsgöl and Hentii mountain ranges, and Ih Khyangan Mountain, though as yet nesting has not been documented. Also an irregular winter visitor in Great Lakes Depression and feeding in planted fruit trees in Uvs and Khovd among other towns across Mongolia.

Japanese Waxwing *Bombycilla japonica* — 16–19 cm

ID Distinguished from very similar Bohemian Waxwing by slightly smaller size/lighter build; black eye-stripe extending back along edge of crest; bright red secondary bar in wings; and bright red (not yellow) terminal tail-band. **Voice** Call similar to that of Bohemian Waxwing but briefer and higher-pitched. **Habitat** Nests in coniferous forest, especially cedar and larch; migratory flocks visit fruit trees in winter. **Behaviour** Similar to Bohemian Waxwing. **Status** Vagrant. One adult photographed in Khalkh River valley, Dornod province, 2 June 2014. Also an unconfirmed record from the Nömrög River valley in 2002. **Conservation** Near Threatened globally due to small population size, habitat loss and persecution in cage bird trade.

Flavescent Bulbul *Pycnonotus flavescens* — 20 cm

ID A slender, thrush-sized bird, mainly olive-green with grey, slightly crested head, streaked breast and upright posture. Unlikely to be confused with any other species in Mongolia. Imm is duller and browner with pale, not dark, bill. **Voice** Song has been described as 'jolly and quick': *joi whiti-whiti-wit*, etc. (Robson 2009); calls are various harsh buzzing notes. **Habitat** Open forest and forest edges. **Behaviour** Insectivorous; gleans branches and flycatches **Status** Vagrant. One in planted poplar trees at Khanbogd, Ömnögobi province, 3 September 1978.

White-throated Dipper *Cinclus cinclus* — 17–20 cm

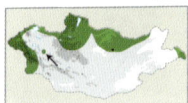

ID Unmistakable due to its chunky form and generally dark brown coloration with sharply contrasting white throat and upper breast. Race *C. c. leucogaster* also has a white belly. Dark morph of *C. c. baicalensis* is reminiscent of extralimital Brown Dipper *C. pallasii* but throat and breast paler. Always closely associated with water. **Voice** Call is a loud, metallic *klink*; also a short *dzit*; song a jumble of harsh squeaky and chirping notes as well as liquid warbles; sings even in winter. **Habitat** Breeds near rivers in high mountains, taiga forest and forest steppe. Winters at lower elevations, where it finds open water with abundant aquatic life. **Behaviour** A remarkable, entertaining, water-loving songbird that dives into fast-moving rivers in search of aquatic invertebrates. Walks underwater along the riverbed and often floats or swims on surface. Builds a domed nest on riverside rock ledges. **Status** Uncommon to fairly common resident breeder and winter visitor across Mongolia, from Mongol-Altai mountain range in the west through Hövsgöl, Khangai and Hentii ranges to the Nömrög River in the far east.

Eurasian Wren *Troglodytes troglodytes* — 9–10 cm

ID Combination of tiny size, compact form, rich brown coloration, short, perennially cocked tail and hyperactivity is highly distinctive. However, compare *Locustella*, *Cettia* and Dusky Warblers. **Voice** Calls include a hard *tac* and chattering *tec-tec-tec*. Song is long, high-pitched and surprisingly loud, including trills and rapid 'machine-gun' rattles. **Habitat** Nests in mature forest with thick litter and undergrowth. Migrants forage in thickets near water. **Behaviour** Quite skulking, though often responds to pishing and squeaking noises. Forages for spiders and other invertebrate prey on ground and in low vegetation. **Status** Rare and local breeding visitor and passage migrant to Hentii range and Buir Lake-Khalkh River-Khyangan Region, late April to early September. [Alt: Northern Wren]

Bohemian Waxwing

adult

adult

adult

juvenile

adult

juvenile

Japanese Waxwing

adult

Flavescent Bulbul

adult

adult

leucogaster

adult

tianschanicus

Eurasian Wren

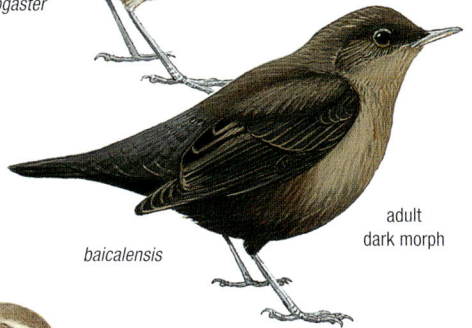

adult

baicalensis

adult
dark morph

White-throated Dipper

adult

dauricus

Marsh Tit *Poecile palustris*   11–12 cm

ID Very similar to Willow Tit from which best distinguished by: restricted and rather tidy (vs. more extensive and irregular) throat bib; glossy (vs. dull) black cap; and absence of pale panel in folded wing (secondaries). However, all of these characters are variable and can be hard to discern. See also voice, distribution and habitat, below. **Voice** Usual call an explosive, sneezing *pitchoo*; song sweet but monotonous *chip-chip-chip* or *tew-tew-tew*. **Habitat** Coniferous and especially deciduous and mixed woodland, as well as wooded gardens and parks in towns. **Behaviour** Cavity nester. Agile and restless; forages in mixed flocks in winter. Dominates Willow Tit where they both breed. **Status** Common resident breeder in northern Mongolia from Uvs Depression through Hövsgöl, Khangai and Hentii ranges, with a disjunct population in eastern Dornod. **Taxonomy** Formerly placed in genus *Parus*.

Willow Tit *Poecile montanus*   11–12 cm

ID Very like Marsh Tit, but differs by slightly larger head, more extensive black bib and pale wing panel. See Marsh Tit for other plumage distinctions. See also habitat, distribution and (especially) voice, below. **Voice** Usual call is a harsh, nasal *djeee r-djeer-djeer*. Song resembles slow version of Wood Warbler's song: *tiu tiu tiu*. **Habitat** Deciduous, coniferous and mixed forest as well as wooded parks and gardens in towns. **Behaviour** Similar to Marsh Tit. Excavates its own nest cavity from decaying wood. **Status** Common resident breeder with more extensive range than Marsh Tit, extending across northern forested areas of Mongolia. **Taxonomy** Formerly placed in genus *Parus*.

Great Tit *Parus major*   13–15 cm

ID The largest Mongolian tit with unmistakable plumage pattern consisting of black 'cowl' and belly stripe; contrasting white cheek patch; greenish back; and a single white wing-bar. Adult male has broader black stripe along lower breast and belly than female. Juv duller, with yellowish cheeks and reduced or absent belly stripe. **Voice** A bewildering variety of loud ringing calls. Song usually a repeated two-note phrase, often rendered as *tea-cher tea-cher tea-cher*. **Habitat** All forest types as well as wooded parks and gardens. **Behaviour** Lively, agile and bold; dominates other tit species in mixed flocks. **Status** Common and widespread resident breeder in forested regions from Mongol-Altai range to easternmost Dornod.

'Turkestan Tit' *Parus* (*major*) *bokharensis*   14–15 cm

ID Plumage pattern very similar to Great Tit, but underparts white, not yellow, and back grey, not green. **Voice** Similar to Great Tit. **Habitat** Nests and winters in willow, poplar and other deciduous woodlands in Bulgan River valley, in the west. **Behaviour** Similar to Great Tit. **Status** Rare and local resident breeder in Bulgan River valley, Khovd province. **Taxonomy** Considered by IOC and BirdLife International to be a subspecies of Great Tit *P. major*.

adult

Marsh Tit

uralensis

adult

baicalensis

adult

Willow Tit

♂

Great Tit

♂

juvenile

'Turkestan Tit'

Siberian Tit *Poecile cinctus*   13–14 cm

ID Similar to Willow and Marsh Tits, but readily distinguished by dull blackish-brown (vs. black) cap, browner back and rusty-brown flanks. Less dapper than Willow or Marsh Tits with fluffy, untidy appearance. Like Willow Tit has extensive black bib. **Voice** Song is a loud and throaty repeated two-syllable phrase *cheeoo-cheeoo-cheeoo*. Call very similar to Willow Tit's. **Habitat** Prefers conifer forest in mountains but also patches of birch and aspen. Moves down to lower edge of conifer forest and rarely to river valleys and wooded parks in winter. **Behaviour** Similar to Willow and Marsh Tits. **Status** Rare and local resident breeder in main Khangai, Hövsgöl and Hentii ranges, with disjunct population in Buir Lake-Khalkh River-Khyangan Region. **Taxonomy** Formerly placed in genus *Parus*. [Alt: Grey-headed Chickadee]

Coal Tit *Periparus ater*   11–12 cm

ID Smallest of Mongolian tits. Plumage pattern similar to Marsh and Willow, but easily distinguished by blue-grey back, two white wing-bars and distinctive white nape-patch. **Voice** High, plaintive *tsee* or *psit*, singly or in series. Song is reminiscent of higher-pitched and speeded-up Great Tit's *pitchee-pitchee-pitchee*. **Habitat** Mainly coniferous forest but also typical taiga birch and aspen copses and wooded parks and gardens. **Behaviour** Generally shyer than other tits, but joins mixed-species flocks in winter. **Status** Fairly common resident breeder across northern Mongolia from Mongol-Altai range and Great Lakes Depression through Khangai, Hövsgöl and Hentii ranges and far eastern Dornod. **Taxonomy** Formerly placed in genus *Parus*.

Azure Tit *Cyanistes cyanus*   12–13 cm

ID An elegant blue-and-white tit, impossible to confuse with any other Mongolian species. **Voice** Calls are varied combinations of rolling buzzy phrases e.g. *tsee-tsee-dzc dzc*. Song is typically a more elaborate version of calls. **Habitat** Willow copses in river valleys and other thickets and woodland edges near water; also parks and garden, especially in winter. **Behaviour** Similar to Marsh and Willow Tits. **Status** Fairly common resident breeder with a broad range extending from Mongol-Altai range and Great Lakes Depression in the west, through Hövsgöl, Khangai and Hentii ranges to the Khalkh-Khyangan Region of far east. **Taxonomy** Formerly placed in genus *Parus*.

Common House Martin *Delichon urbicum*   13–15 cm

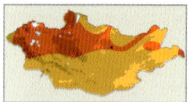

ID Glossy blue-black upperparts and sharply contrasting white underparts (duller in juv) distinguish this species from all other Mongolian swallows except very similar Asian House Martin, which see for points of difference. **Voice** Somewhat toneless twittering *churi-ri churi-ri* and more emphatic *prrrit*. **Habitat** Nests on cliffs and artificial structures near wetlands, open areas and settlements. **Behaviour** Builds a cup-shaped nest of mud pellets under eaves or on suitable ledges, visiting puddles to collect mud. **Status** Fairly common breeding visitor across the country apart from driest regions of extreme south, and fairly common passage migrant throughout, early May to late August (early September in the south). [Alt: Northern House Martin]

Asian House Martin *Delichon dasypus*   13 cm

ID Very similar to Common House Martin, from which best distinguished in flight by less extensive and finely streaked white rump; dirty greyish tinge to underparts (vs. clear white); dusky (vs. white) undertail-coverts; and (especially) dark vs. pale underwing-coverts. **Voice** Similar to Common House Martin. **Habitat** As Common House Martin, though nests at cliff sites in Mongolia. **Behaviour** Similar to Common House Martin. **Status** Very rare and local breeding visitor, reported to nest in Khoridol Saridag and Bayany Nuruu in Hövsgöl Mountains. Passage dates presumably as Common House Martin.

Siberian Tit

adult

adult

Coal Tit

adult

yenisseensis

Azure Tit

Common House Martin

adult

adult

Asian House Martin

Sand Martin *Riparia riparia*   12–13 cm

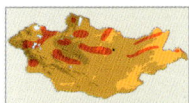

ID Distinguishable from other brown Mongolian swallows by sharply defined dark breast-band and by habitat; compare Pale Martin. **Voice** A dry *chrrr chrrr* flight call. **Habitat** Nests in sandy banks of rivers and lakes, but occurs countrywide, including arid steppe and desert, on migration. **Behaviour** Breeds in dense colonies, excavating its own nesting tunnels in sand banks and sand quarries. **Status** Locally common breeding visitor across northern Mongolia where suitable nesting habitat exists and uncommon passage migrant throughout, late April to late August. [Alt: Common Sand Martin, Bank Swallow]

Pale Martin *Riparia diluta*   11–13 cm

ID Distinguished from very similar Sand Martin by paler and greyer upperparts, dingier underparts and diffuse breast-band. Juv has pale feather fringes on upperparts. **Voice** Toneless hard twittering and a brief *ret* or *brret*. **Habitat** As Sand Martin. **Behaviour** Nests colonially in sandy banks near lake and river shores. Roosts in reedbeds and on artificial structures near water. **Status** Uncommon breeding visitor and very uncommon passage migrant with confirmed records from Lake Khar-Us, Khovd province, Ulz and Khalkh river valleys, Dornod province, Lake Khunt, Bulgan province and Tuul river valley in central Mongolia. The species may be overlooked because of its similarity to Sand Martin, especially where the two species co-occur. Presumably present late April to late August. **Taxonomy** Considered to be a subspecies of Sand Martin *R. r. diluta* by BirdLife International and other authorities. [Alt: Pale Sand Martin]

Eurasian Crag Martin *Ptyonoprogne rupestris*   14–15 cm

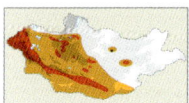

ID When perched, best distinguished from Mongolia's two other brown swallows by finely streaked throat and generally darker ('dirty') underparts with no trace of breast-band. However, best recognised in flight by white spots near end of frequently fanned tail, and by contrasting blackish underwing-coverts. Habitat also different. **Voice** Calls are various chips and chirrs; also a note described as finch-like or lark-like. Song a soft rapid twittering. **Habitat** Breeds on rocky cliffs in high mountains, especially deep gorges; rarely on buildings or in walls. **Behaviour** Builds a half cup of mud attached to rock faces in caves and crevices. Flight heavier and more direct, less fluttering, than other hirundines. Habitually patrols the edges of ravines, but also feeds on insects at height. **Status** Locally common breeding visitor throughout Altai mountain range, including Mongol and Gobi-Altai and Khangai, and uncommon passage migrant through breeding areas, Hentii and Gobi, early May to late August.

Barn Swallow *Hirundo rustica*   17–21 cm

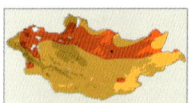

ID Could be confused with Red-rumped Swallow, which also has glossy blue-black upperparts and long forked tail. However, readily distinguished by dark throat, unstreaked breast and belly, and absence of reddish rump. Juv is duller and paler on forehead, throat and underparts. Two distinct subspecies occur in Mongolia: *H. r. tytleri* has deep rufous breast; *H. r. rustica* has a whitish breast. **Voice** Song is a pleasant, rapid twittering; in alarm, gives a sharp *vit* or *veet*. **Habitat** Nests on artificial structures such as bridges and interiors of unused buildings in all open habitats. **Behaviour** Builds an open cup nest of mud mixed with straw or other plant material. Frequently flies very close to the ground, often feasting on insects around livestock herds. **Status** Common breeding visitor and passage migrant throughout Mongolia, late April to late August (early September in the Gobi).

Red-rumped Swallow *Cecropis daurica*   14–19 cm

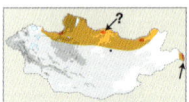

ID Distinguished from similar Barn Swallow by rusty face and collar; pale throat; buffy, streaked breast and belly; black undertail-coverts; and reddish rump. Juv is duller and browner than adult, with paler collar and rump, and shorter tail-feathers. **Voice** Song resembles Barn Swallow, but less sweet and lower-pitched; alarm call a rolling *kiiir*. **Habitat** Open areas near cliffs, riverbanks and human habitation. **Behaviour** Builds a closed nest of mud and plant material with an entrance tunnel, and attaches it to natural or artificial ledges. **Status** Rare breeding visitor in western Hövsgöl province and eastern Dornod and uncommon passage migrant through breeding range and southward, late April to late August.

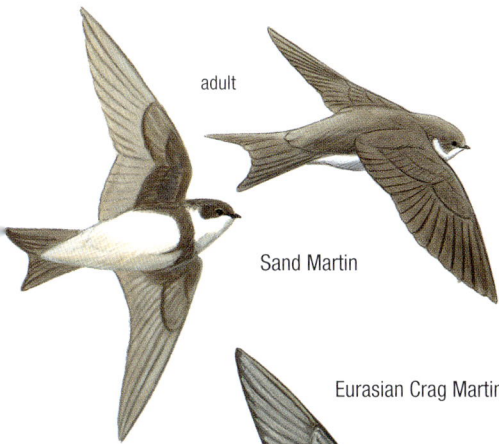

adult

Sand Martin

Pale Martin

Eurasian Crag Martin

adult

adult

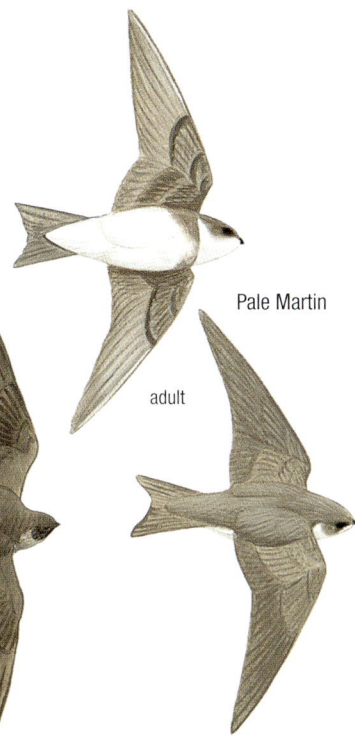

adult

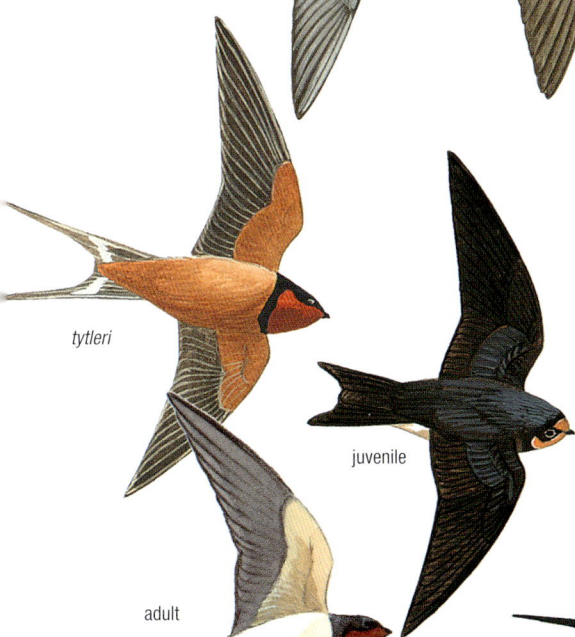

Red-rumped Swallow

tytleri

juvenile

adult

adult

rustica

daurica

Barn Swallow

Crested Lark *Galerida cristata* 17–19 cm

ID Larger and stockier than Eurasian Skylark, with longer decurved bill and (most notably) a much longer and more pointed crest. It also lacks Eurasian Skylark's white outer tail-feathers and pale trailing edge on secondaries. Compare also distribution. **Voice** Sings from ground and during display flight, a pleasant and varied series of twittering notes. Call is a plaintive trisyllabic *tree-wee-woo*, also *too-wee*. **Habitat** Low-lying and sparsely vegetated areas with patchy tall grass in transit zone between arid steppe and Gobi Desert. Sometimes occurs in town centres and settlements in drier areas. **Behaviour** Feeds on ground, shuffling along as it searches for seeds and insects. **Status** Uncommon and local resident breeder and partial migrant from Great Lakes Depression and Mongol-Altai, south and east in drier areas through the Gobi. Generally absent from more humid areas to north and the far east, though wandering birds appear to the north, including near Ulaanbaatar. Birds present in breeding areas mid-April to early October; degree of wintering depends on snow cover and seed availability.

Mongolian Short-toed Lark *Calandrella dukhunensis* 14–16 cm

ID Short-toed lark species may be distinguished from other lark species by their smaller size, overall paler/greyer coloration and short, thick finch-like bills. Mongolian Short-toed (only slightly larger) is best identified by clear breast (which may have minor streaking at upper margins), long tertials reaching tip of wing, and in most individuals small black patches on upper breast (like broken collar). It also may show a contrasting dark wing-bar. Asian Short-toed Lark is very similar, but has finely, but distinctly, streaked breast and shorter bill than Mongolian. See also voice, habitat and distribution. **Voice** Song is a halting and rather toneless twittering; also a longer song confusable with other larks. Calls include a chirping *tchrip* also drier *prrrt*; at times very similar to Asian. **Habitat** Well-vegetated areas with scattered small shrubs and tall grasses in dry steppe and Gobi Desert, as well as valleys of rivers and lakes. **Behaviour** Nests and feeds on ground. Forms post-breeding flocks of up to 500, mixing with other lark species. **Status** Fairly common breeding visitor in eastern steppe (*dukhunensis*) and some areas of north-west Mongolia (*longipennis*); fairly common passage migrant throughout. Generally present mid-April to early October; a few individuals may winter in steppe, depending on snow cover and food availability. Race *orientalis* may occur in northern Mongolia. **Taxonomy** Formerly considered to be a subspecies of Greater Short-toed Lark *C. brachydactyla*.

Asian Short-toed Lark *Alaudala cheleensis* 13 cm

ID Best distinguished from Mongolian Short-toed Lark by finely, but distinctly streaked breast, shorter, stubbier bill, and absence of black patches at sides of upper breast. At very close range the tips of tertials never reach the tips of primaries in the folded wing (Mongolian has very long tertials). See also Voice. **Voice** Song is sweeter and more varied than that of Mongolian Short-toed Lark, including some mimicry. Flight call quite different from Mongolian – is a dry, gravelly *drrrt* sometimes compared to Sand Martin's call. **Habitat** All types of steppe with tall grasses and small shrubs (e.g. *Caragana*) including drier areas in valleys of saline lakes and marshes. **Behaviour** Similar to Mongolian Short-toed Lark. **Status** Common and local breeding visitor and passage migrant throughout, late March to early April and early September to early October. **Taxonomy** Recent field research has determined that two *Alaudala* larks occur in Mongolia, *A. cheleensis* (Asian Short-toed Lark) in the north-east and another, very similar, form in the south which is currently considered to be a subspecies of Lesser Short-toed Lark *A. rufescens heinei*; the taxonomy of the latter will be the subject of a forthcoming paper.

Eurasian Skylark *Alauda arvensis* 16–18 cm

ID Shorter crest, white outer tail-feathers and white trailing edge of wings are main points of difference from Crested Lark. Larger, with more variegated upperparts and longer, more pointed bill than short-toed lark species. Juv has less contrasting plumage, pale-tipped buff upperparts and buff-edged flight- and tail-feathers. **Voice** Song is a continuous exuberant medley of high twitters, chirps and whistles, frequently delivered in impressive towering song-flight and lasting as long as 15 minutes. Common call is a rolling *chrrik* or *chirrrup*. **Habitat** Variety of open, typically moist habitats in steppe, valleys of lakes and rivers, and cultivated land. **Behaviour** Feeds and nests on ground. Small numbers of adult birds flock with other larks in winter, depending on snow cover and seed availability. **Status** Generally common breeding visitor and passage migrant, absent only from highest and driest areas; occurs throughout country on migration. Mainly mid-April to early September; sporadic in winter.

adult

juvenile

adult

Crested Lark

juvenile

adult

adult

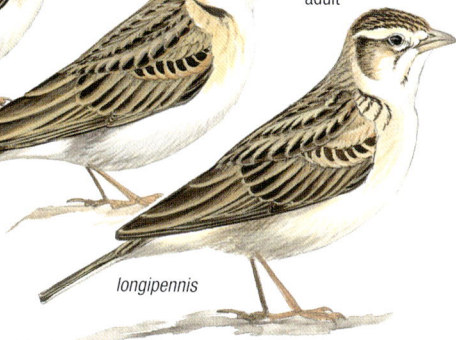

dukhunensis

longipennis

adult

Mongolian Short-toed Lark

adult

cheleensis

heinei

adult

Asian Short-toed Lark

adult

adult

juvenile

Eurasian Skylark

Horned Lark *Eremophila alpestris*

16–19 cm

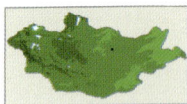

ID Black-and-yellowish head pattern unique and unmistakable. Female and non-breeding birds less strongly marked with reduced or absent 'horns', but still highly distinctive. Juv is heavily spotted black, white and tan, and shows trace of adult's moustache. **Voice** Song combines high trills and chuckling phrases with harder chattering sounds. Calls are soft and weak, *seee* or *seee-tu*. **Habitat** Occurs in all open habitats, including subalpine meadows and desert as well as all forms of steppe. **Behaviour** Forms post-breeding flocks of up to 1,500 individuals, and in winter joins other larks and seed-eaters to feed on snow-free roadsides. **Status** Very common to locally abundant resident breeder and common winter visitor throughout. [Alt: Shore Lark]

Mongolian Lark *Melanocorypha mongolica*

18 cm

ID This large, stocky, heavy-billed lark with striking rufous, black and white plumage pattern is unlike any other lark species in Mongolia, except for vagrant White-winged Lark. The latter has slighter build, smaller bill and lacks Mongolian Lark's prominent black throat collar. Adult female is duller and more streaked above. Juv lacks reddish-brown in plumage and is scalier above. **Voice** Prolonged twittering flight song is slightly deeper and richer than Eurasian Skylark's; also various harsh and shrill calls. **Habitat** All forms of steppe, preferring relatively rich grasslands, but also nests in sparsely vegetated dry habitats with patchy tall grass. **Behaviour** Often nests behind dried horse and cattle droppings, grass tussocks and short shrubs. Forms post-breeding flocks of up to 2,500 individuals, and in winter feeds in snowless areas with Horned Larks and Père David's Snowfinches. **Status** Very common resident breeder and common passage migrant throughout Mongolia except for taiga and alpine zones.

White-winged Lark *Alauda leucoptera*

17–19 cm

ID Adult male is similar to much commoner Mongolian Lark, but slighter, with smaller bill, greyer underwing and absence of black patches on upper breast. Adult female and juv lack rusty crown and cheek patch and have overall streaky appearance. **Voice** Song and calls similar to Eurasian Skylark, but more melodious. **Habitat** Dry areas with tall grasses in valleys of salty lakes. **Behaviour** Similar to Mongolian Lark. **Status** Vagrant. Recorded in open dry steppe near Lake Erkhil and Khatgal soum, Hövsgöl province, and Lake Telmen, Zavkhan province, in February and March. **Taxonomy** Formerly placed in the genus *Melanocorypha*.

Black Lark *Melanocorypha yeltoniensis*

18–20.5 cm

ID Adult male is unmistakable with combination of large size and stout proportions; all-black plumage (with grey feather edgings in autumn); and heavy, pale bill. Adult female is a rather anonymous chunky grey bird, but heavy bill and especially blackish wings (including underwing-coverts) are distinctive. **Voice** Song is similar to Eurasian Skylark but more rapid. Calls also like that species but more trilling and buzzing. **Habitat** Steppes, often near marshes and saline areas. **Behaviour** Similar to Mongolian Lark. **Status** Vagrant. One male near Lake Khar-Us, Khovd province, 14 February 1980.

brandti

♂

♂

flava

Horned Lark

adult

adult

juvenile

Mongolian Lark

♂

♀

♂

juvenile

White-winged Lark

juvenile

♀

♀

♂

Black Lark

Beijing Babbler *Rhopophilus pekinensis* 17 cm

ID Resembles a long-tailed *Locustella* warbler or small babbler, best distinguished by strong black moustachial stripe, rufous streaks on flanks and pale iris. Despite its alternative name, does not have a white brow. **Voice** Song is a sweet, inflected *dear-dear-dear-dear*; call a pleasant, *cheee-anh*. **Habitat** Dry, rocky hillsides with scattered shrubs. **Behaviour** Skulking and difficult to see; runs with tail cocked. **Status** Vagrant. Two birds recorded at Shar Khuls oasis, Trans-Altai Gobi, July and August 1943. **Taxonomy** Recent research and analysis (Gelang *et al.* 2009; Alström *et al.* 2013) refer this species along with other warblers to the babbler family (Timaliidae). [Alt: White-browed Chinese Warbler, Chinese Hill Warbler, Chinese Hill Babbler].

Common Grasshopper Warbler *Locustella naevia* 12–14 cm

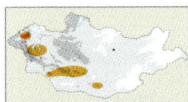

ID Similar to Lanceolated and Pallas's Grasshopper Warblers, but adult lacks bold breast streaking of former and colourful, contrasting upperparts of latter. Imm from imm Lanceolated by minimal streaking below and from imm Pallas's by latter's unstreaked rump and tail. Occurs in three different colour morphs (in adult and juvenile). **Voice** Song is a continuous, dry, toneless, insect-like trill like an angler's reel spinning, very similar to Savi's but slower and not as high-pitched. Call is a high, dry *tick* or *tsit* singly or in pairs. **Habitat** Nests in dense shrubbery and sedge tussocks bordering rivers and lakes, but also margins of upland forest clearings; seldom uses reedbeds. **Behaviour** Shy and mouse-like, rarely seen until flushed or when male sings from exposed perch. **Status** Rare and very local breeding visitor, with confirmed nesting only in valley of Lake Achit in Mongol-Altai range. Breeding suspected at Buyant River near Khovd town and possibly near other lakes in Great Lakes Depression and Valley of the Lakes. Rarely recorded in suitable habitat near springs or lakes in Gobi Desert and in breeding areas during migration. Late April to late August (early September in the Gobi).

Lanceolated Warbler *Locustella lanceolata* 11–12 cm

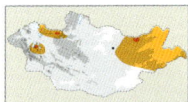

ID The smallest *Locustella* warbler. Adult distinguished from similar Common and Pallas's Grasshopper Warblers, by distinct black streaks and tear-drop spotting on breast and undertail-coverts. Less boldly streaked imm similar to imm Common, which can have streaked upper breast, but lacks side streaking. Pallas's has contrasting rusty rump and dark, white-tipped tail. **Voice** Song is a cricket-like 'reeling', similar to that of Common Grasshopper and Savi's Warblers, but higher-pitched and given in shorter bursts. Calls include a raspy, repeated *tsuh-tsuh-tsuh* in varied cadence; also, a softer *tuck, tuck…* and *chit*. **Habitat** Breeds in tussocky sedge meadows, reedbeds and shrubby borders of other wetlands, as well as edges of damp forest clearings. Visits drier, less dense habitats on migration. **Behaviour** A shy skulker like other grasshopper warblers, easier to see during migration than when nesting. **Status** Very rare and local breeding visitor in widely scattered areas of both western and eastern Mongolia, including Tes River valley (northern Uvs Depression); lakes Khar and Khar-Us; and possibly along Onon and Balj Rivers. Scare passage migrant, only rarely recorded away from breeding areas, late April to early September.

Savi's Warbler *Locustella luscinioides* 13–15 cm

ID Lack of streaking and generally warm-brown coloration distinguish this from other Mongolian *Locustella* warblers. From similar reed warblers, with which it shares habitat preference, by rounded, broad-based tail, very long undertail-coverts and song. **Voice** Song is a high, dry, cricket-like 'reel'. Call is a sharp *tchipp!* **Habitat** Prefers to nest in extensive reedbeds, rarely in other dense wetland vegetation. **Behaviour** Not as shy as other members of its genus. Males sing from prominent perches. Nest placed low in cover. **Status** Recorded as rare and local breeding visitor and rare passage migrant only in valleys of lakes Khar and Khar-Us in Great Lakes Depression, late April to late August.

Beijing Babbler

adult

adult
brown morph

Common Grasshopper Warbler

adult
yellow morph

adult
grey morph

juvenile
yellow morph

adult

Lanceolated Warbler

juvenile

adult

fusca

Savi's Warbler

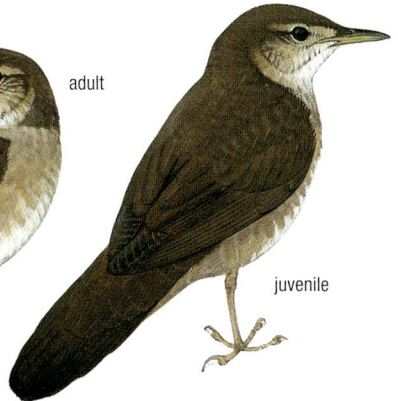

juvenile

Marsh Grassbird *Helopsaltes pryeri*   13 cm

ID In Mongolia, most likely to be mistaken for Pallas's or Common Grasshopper Warblers due to bold streaking on upperparts, longer tail, but rounder-headed and smaller-billed than those species, and lacks any trace of streaking below. In general, resembles long-tailed *Cisticola*. See also Behaviour. **Voice** Song is an extended, slightly liquid, somewhat musical twitter. Calls include brief, dry *juk juk juk, chak chak*, and more drawn-out, rising *trrrrrik*. **Habitat** Strongly associated with reedbeds and tall sedges/grasses. **Behaviour** Climbs and balances in reeds and performs song flight; not especially skulking. **Status** Possibly very rare and local breeding visitor in Hentii and Dornod provinces; males singing in reedbeds at Lake Tashgain Tavan, Dornod province, in June and July in the last decade. **Conservation** Considered Near Threatened globally due to small, isolated populations globally and destruction of its wetland habitat. **Taxonomy** Formerly placed in genus *Megalurus* or *Locustella*. [Alt: Japanese Swamp Warbler]

Pallas's Grasshopper Warbler *Helopsaltes certhiola*   13–14 cm

ID The most distinctive and common *Locustella* warbler in Mongolia with a diagnostic combination of small size; grey, streaked crown; distinct whitish supercilium; rufous upperparts boldly streaked and spotted in black and white; and fairly short dark tail with white tips contrasting with rufous rump. Compare Lanceolated and Common Grasshopper Warblers. **Voice** Song recalls *Acrocephalus* warblers, a highly varied series of trills, chirps, whistles and rattles. Calls include a sharp *chipp!* and a dry rattle: *trrrrrr*. **Habitat** Nests in sedge tussocks, scrubby vegetation and reeds at edges of lakes, bogs and other wetlands. Inhabits many other habitats on migration including steppe and desert, gardens and parks. **Behaviour** Very shy when nesting, creeping mouse-like through deep cover; more easily seen on migration where dense cover is scarce. **Status** Fairly common breeding visitor in all major marshland ranges across northern Mongolia and common passage migrant throughout, late April to late August (early September in the Gobi). **Taxonomy** Formerly placed in genus *Locustella*. Two subspecies occur in Mongolia, *H. c. centralasiae* and *H. c. sparsimstriatus*.

Middendorff's Grasshopper Warbler *Helopsaltes ochotensis*   16 cm

ID In Mongolia, most likely to be confused with Gray's Grasshopper Warbler, but lacks that species' grey breast and has a distinct whitish supercilium set off by a dark eye-stripe. Compare also Savi's and *Acrocephalus* reed warblers. **Voice** Song is a rich, 'grinding' *Weechi-weechu-weechi-chuchu* and variations. Call is a dry *it-it-it*. **Habitat** Wet meadows with thick grasses, reedbeds and dense willow copses and other shrubbery on wetland edges. **Behaviour** Shy and skulking. **Status** Vagrant. One collected near Lake Bayan of Khalkh River, Dornod province, 14 July 1974. **Taxonomy** Formerly placed in genus *Locustella*.

Gray's Grasshopper Warbler *Helopsaltes fasciolatus*   16.5–18 cm

ID Adult is best distinguished from other unstreaked *Locustella* warblers by large size, dark olive-brown upperparts and (especially) grey breast, contrasting with whitish throat and belly. **Voice** Song is a repeated short, rich, warbled phrase: *chupichakachaka*. Calls include a trilled *cherr-cherr* and a clacking *tchdek*. **Habitat** Rarely nests in wetlands, preferring tall, dense tangles of grasses or shrubs at forest edges and in clearings. **Behaviour** Has been described as 'master skulker', even singing from dense cover and prefering to run on the ground like mouse rather than fly. **Status** May breed in two widely separated parts of country: Eg River basin and Darkhad Depression in Hövsgöl Mountains; and Khalkh River valley, Dornod province, and rare passage migrant within breeding range. Presumably present late April to late August. **Taxonomy** Formerly placed in genus *Locustella*.

adult

Marsh Grassbird

juvenile

centralasiae

adult

sparsimstriatus

adult

Pallas's Grasshopper Warbler

1st-winter

adult

Middendorff's Grasshopper Warbler

1st-winter

adult

Gray's Grasshopper Warbler

PLATE 75: BUSH WARBLERS & REED WARBLERS I

Baikal Bush Warbler *Locustella davidi*   14 cm

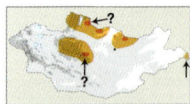

ID Like stubby *Locustella* or *Acrocephalus* warbler, best distinguished from other dull brown species by fine necklace of streaks (sometimes faint) beneath white throat, by barred undertail-coverts and by song. **Voice** Song is a loud, fast, high-pitched, rapidly repeated 3–4 note phrase. Call is a quiet *tchik*, or *tchit*. **Habitat** Dense thickets and scrub in open grassy areas, e.g. at edge of mountain taiga. **Behaviour** Skulking and shy; runs in dense vegetation like mouse. **Status** Very rare and local breeding visitor in Khangai, Hövsgöl, Hentii and Ih Khyangan mountain ranges, and very rare passage migrant within breeding range, late April to late August. **Taxonomy** Until recently, considered a race of Spotted Bush Warbler *B. thoracicus*. **Taxonomy** Formerly placed in genus *Bradypterus*. [Alt: David's Bush Warbler]

Chinese Bush Warbler *Locustella tacsanowskia*   14 cm

ID Similar to Baikal Bush Warbler, but paler olive-brown (vs. rich brown) above and with distinctly yellowish-buff tone below. Black bill distinct in adult. Breast spotting and undertail barring much less pronounced than in Baikal. See also voice and distribution. **Voice** Song is a dry cicada-like monotone buzz: *dzeeeezee*. Call is a low *churr churr*. **Habitat** Dense scrubby thickets in mountain meadows and valleys near forest edges. **Behaviour** Shy and skulking. **Status** Uncommon breeding visitor and passage migrant, known mainly from river valleys in the Hentii mountain range; probably also nests in Buir Lake-Khalkh River-Khyangan Region in far east, late April to late August (early September in the Gobi). **Taxonomy** Formerly placed in genus *Bradypterus*.

Booted Warbler *Iduna caligata*   11–12 cm

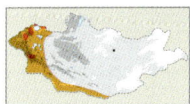

ID May be confused with Blyth's Reed and Paddyfield Warblers but has more delicate build; more rounded head; shorter, thinner bill with dark tip on yellow lower mandible; shorter tail; and dark feet contrasting with pinkish legs. Upperparts grey-brown (vs. olive or rufous). **Voice** Song is a rapid bubbly mix of warbles and squeaks in repeated distinct phrases. Call is a quiet but grating *tchak*. **Habitat** Prefers to nest in dense thickets, e.g. wild rose and *Caragana* bushes, or willow; less commonly in reedbeds. Uses more diverse shrublands on migration. **Behaviour** Shy and skulking. **Status** Fairly rare and local breeding visitor in western Mongolia, including lake and river valleys in Northern Uvs, Great Lakes and Baruunkhurai Depression (Bulgan River valley). Scarce migrant in oases in Gobi Desert, as well as within breeding areas, late April to late August (early September in the Gobi). **Taxonomy** Formerly placed in genus *Hippolais*.

Blyth's Reed Warbler *Acrocephalus dumetorum*   12–14 cm

ID Similar to Paddyfield Warbler, but differs in greyish-olive (vs. more rufescent) upperparts; less prominent head pattern; lower mandible with diffuse dark tip; and other details described under that species. Compare also Booted Warbler. **Voice** Song is much slower, simpler and more deliberate than other warblers, e.g. *sweet, chiew, chick-chick* with phrases typically repeated several times. Calls include a distinctive hard *chek* and *tchirr*. **Habitat** Nests and migrates in a mixture of low and secondary habitats such as riverine willow or birch copses, brushy clearings in moist forest edges, as well as parks and gardens on migration; only rarely uses reedbeds. **Behaviour** Unlike other *Acrocephalus* warblers, forages mainly in woody vegetation, hopping through foliage with tail held slightly raised. **Status** Very rare and local breeding visitor to Uvs and Khar-Us Lakes in Great Lakes Depression and Bulgan River in Baruunkhurai Depression. No records outside of breeding period, early May to late August.

Paddyfield Warbler *Acrocephalus agricola*   12–13 cm

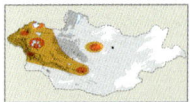

ID A small, rather delicate-looking *Acrocephalus* warbler. Most like Blyth's (with which it co-occurs) but with warmer (vs. greyish) brown coloration above, including contrasting rufous rump and rufous buff (vs. pale olive) underparts; broader pale supercilium with dark (but not black) borders above and below; shorter, finer bill; and comparatively long, rounded tail. Compare also Booted Warbler. **Voice** Song is high-pitched, 'streaming' and more musical than other *Acrocephalus* songs, incorporating much mimicry. Calls include *check, cherrr,* and *zack*. **Habitat** Nests primarily in reedbeds, but rarely in other waterside vegetation; less selective on migration. **Behaviour** Tends to nest and feed low down in reeds. Flicks tail constantly. **Status** Uncommon breeding visitor in riverine and lacustrine wetlands of Mongol-Altai mountain range, Northern Uvs and Great Lakes Depression, Zavkhan and Bogd Rivers and Lake Dashinchilen, Bulgan province. It may also nest in Bulgan River valley, Khovd province, Lake Ögii (Khangai mountain range) and Lake Orog in Valley of the Lakes. Uncommon passage migrant through breeding range, late April to late August.

adult

adult plain

juvenile

adult streaked

Baikal Bush Warbler

Chinese Bush Warbler

juvenile/
1st-winter

adult

Booted Warbler

adult
worn

juvenile/
1st-winter

adult

Blyth's Reed Warbler

Paddyfield Warbler

adult

adult
worn

1st-winter

Black-browed Reed Warbler *Acrocephalus bistrigiceps*   14 cm

ID The most distinctive reed warbler in Mongolia. Small, with a diagnostic face pattern: distinct pale supercilium bordered by black eye-stripe below and black crown-stripe above. **Voice** Typical *Acrocephalus* song, high-pitched and fast with lots of chirps and rattles, but sweeter and not as 'scratchy' as larger species. Calls include a distinctive rasping *chirrr* and short *zeck*. **Habitat** Nests in reedbeds, but also dense shrubby vegetation of rivers and lakes; generally sticks to wetland habitats on migration. **Behaviour** Male sings from prominent perch high in reeds, otherwise rather skulking. **Status** Fairly common breeding visitor at Lake Tashgain Tavan and Khalkh, Guu, Azarga and Herlen River valleys of eastern Mongolia. Uncommon passage migrant in breeding range, late April to late August.

Thick-billed Warbler *Arundinax aedon*   16–17 cm

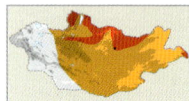

ID Generally very similar to Great and Oriental Reed Warblers, but readily distinguished by much shorter and thicker bill with all-yellow lower mandible lacking black tip; eye-ring and otherwise rather blank face pattern creating an 'innocent' or 'spacey' look; more prominently rounded tail; and short (vs. very long) primary projection. See also habitat and behaviour. **Voice** Song generally similar to Great Reed Warbler – with mix of chatters and warbles, including mimicry – but faster, more slurred, somewhat more musical and not as loud. Call is a low-pitched *chuk* often in a short series; also a harder *chack*. **Habitat** Prefers wooded habitats such as willow copses near taiga bogs or upland thickets and understorey bushes in open forest; not a marsh species. On migration it occurs in many other habitats including steppe, desert and urban parks and gardens. **Behaviour** Stays in dense cover even when singing; movements and flight rather slow and clumsy. Often cocks tail while perched. **Status** Uncommon breeding visitor in Hövsgöl, Khangai and Hentii mountain ranges as well as Darkhad Depression, Orkhon-Selenge River Basin and Buir Lake-Khalkh River-Khyangan Region. Common passage migrant throughout Mongolia, late April to early September. **Taxonomy** Formerly placed in genus *Acrocephalus*.

Clamorous Reed Warbler *Acrocephalus stentoreus*   18–20 cm

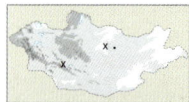

ID Best distinguished from similar Great Reed and Oriental Reed Warblers by notably long and narrow bill and the narrow, but well-defined (vs. broad and diffuse) yellowish supercilium in front of the eye. Also distinguished from Great Reed by short primary projection and strongly rounded tail. **Voice** Song is like Great Reed's but more tentative, with less certain pattern; also softer and more melodious. Calls include harsh, guttural, *chaks* and *gurrrs*. **Habitat** Reedbeds and bushes near wetlands. **Behaviour** Like Great Reed, but male may sing more often from prominent perch. **Status** Vagrant. Known only from two single birds: at Lake Biger, Trans-Altai Gobi, June 1990; and Lun district, Töv province, June 2006.

Great Reed Warbler *Acrocephalus arundinaceus*   16–20 cm

ID A large reed warbler, very similar in most respects, including song, to Oriental Reed Warbler. Longer primary projection of Great is diagnostic but may be difficult to see in the field. Helpfully, ranges of Great Reed (western) and Oriental Reed Warblers (eastern) are not known to overlap in Mongolia (though western extent of Oriental's range is uncertain). See Clamorous Reed Warbler and Thick-billed Warbler for further distinctions. **Voice** Song is deep, loud, guttural and deliberate series of phrases that has been rendered as *karra-karra, keek-keek-keek, kar-kar-kar...*etc. (Baker 1997); common calls are *chack* and *krrrr*. **Habitat** Nests in reedbeds, occasionally foraging in marsh-edge shrubs. Visits other habitats including brushy steppe and desert on migration. **Behaviour** Prefers to keep low in dense reedbeds even when singing, so often difficult to see even when close by. Flight is laboured and jerky. **Status** Uncommon breeding visitor in lacustrine and riverine reedbeds at Lake Achit in Mongol-Altai mountain range, the large water bodies of Great Lakes Depression and possibly Bulgan River in Baruunkhurai Depression. Uncommon passage migrant, mainly through breeding range, late April to late August.

Oriental Reed Warbler *Acrocephalus orientalis*   19 cm

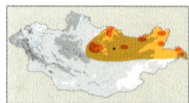

ID Very similar to Great Reed Warbler with which it was once considered conspecific. It is slightly smaller, with a shorter primary projection, and may show streaks on breast, all of which may be difficult to discern in the field. However, it occurs in central and eastern Mongolia, largely separated from range of Great Reed Warbler in the west, though western extent of its range is uncertain. Compare also Thick-billed Warbler and Clamorous Reed Warbler. **Voice** Very similar to Great and Clamorous Reed Warblers, though said by some authors to be richer, softer and less rhythmic. Calls as Great Reed. **Habitat** Nests in reedbeds; may forage in other wetland vegetation. **Behaviour** Similar to Great Reed Warbler. **Status** Uncommon and local breeding visitor in lacustrine and riverine reedbeds in Orkhon-Selenge River Basin, Hentii mountain range and Lake Buir-Khalkh River-Khyangan Region in the east, and uncommon passage migrant, early May to late August.

1st-winter

Black-browed Reed Warbler

adult

adult

Thick-billed Warbler

adult

adult

Clamorous Reed Warbler

adult

Great Reed Warbler

1st-winter

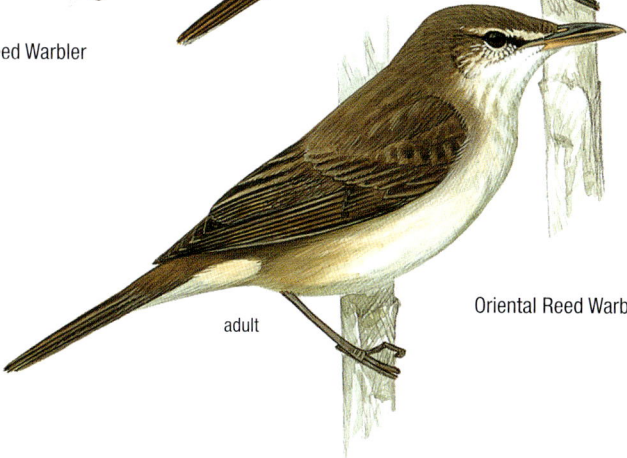

adult

Oriental Reed Warbler

Dusky Warbler *Phylloscopus fuscatus*   10.5–12 cm

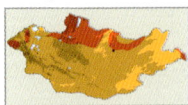

ID A distinctly dull brown *Phylloscopus* with buffy underparts, closest in appearance to much rarer Radde's Warbler, which see for distinctions. Siberian Chiffchaff differs in grey-brown plumage lacking green tinge on wings and tail; less distinct supercillium and eye-stripe; black (vs. reddish-brown) legs; habitat; and behaviour. **Voice** Song is a simple, repeating short phrase, e.g. *chewee chewee chewee chewee*, varying in tone. Call is a distinctive low, abrupt *chek*. **Habitat** Breeds in mountain taiga scrub and willow thickets near bogs and other wetlands. Seeks shrubby cover elsewhere in steppe and desert, parks and gardens on migration. **Behaviour** Usually forages low in undergrowth; flicks wings and tail often. Rather shy and skulking, though readily called out. **Status** Fairly common breeding visitor across all northern mountain ranges from Great Lakes Depression to Dornod, and common passage migrant throughout, mid-April to mid-September.

Sulphur-bellied Warbler *Phylloscopus griseolus*   11 cm

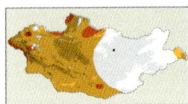

ID Readily distinguished from other brownish *Phylloscopus* by combination of long yellowish ('sulphury') supercilium (colour most vivid in front of eye); distinct dull yellow wash on lower breast and belly; and mostly yellow bill. See also habitat and distribution. **Voice** Song is a simple series of about six high *tsee-tsee-tsee…* notes, sometimes starting with flat, lip-smacking note and rising towards the end. Call is a soft *quip* or *pick*. **Habitat** Breeds on dry rocky slopes with scattered juniper, *Caragana* or other shrubs at high altitude (2,600–4,000 m). Uses other, mainly brushy habitats at lower altitudes on migration. **Behaviour** Forages mainly on or close to ground, gleaning insects from vegetation. **Status** Fairly rare and local breeding visitor in alpine zone of mountains in Great Lakes Depression, and Mongol-Altai, Gobi-Altai and Hövsgöl ranges. Rare passage migrant within its breeding range and in far east, late April to late August (early September in the Gobi).

Yellow-streaked Warbler *Phylloscopus armandii*   12 cm

ID Very similar to Radde's Warbler, but with slightly more delicate build, somewhat finer bill, and thinner legs. **Voice** Song is similar to Radde's, but 'weaker'. Call is a sharp, bunting-like *tzic*, but also a *tack* call similar to that of Dusky Warbler. **Habitat** Breeds in montane willow and poplar groves and tall thickets as well as spruce stands at 1,400–3,500 m in NE and C China. Forages in other types of woody vegetation on migration. **Behaviour** Shy while nesting, foraging in low, dense vegetation. **Status** Vagrant. Two records: two singing males reported from Yolyn Am, Gurvansaikan range, Ömnögobi province, 13 June 2000; and one captured in mixed forest with willow, poplar, larch and fruit trees in Khalkh River valley, Dornod province, 25 August 2010.

Radde's Warbler *Phylloscopus schwarzi*   11.5–12.5 cm

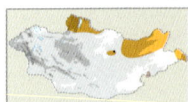

ID Similar to Dusky Warbler in overall brown and buff coloration, but differs by chunkier build; notably shorter, stouter bill; broader supercilium that is palest at rear and more diffuse in front (vs. the opposite in Dusky); distinctly pale and rather thick legs; and orange-buff (vs. uniform pale buff) undertail. Compare also Yellow-streaked Warbler. **Voice** Song is a series of rich, emphatic, sometimes bubbly phrases. Call is a low *chrep*, softer and 'looser' than Dusky. **Habitat** Typically forages in shrubby vegetation or low in small trees, but to be looked for in any wooded or brushy habitats in company of other warblers in migration. **Behaviour** Keeps to cover, but more sluggish in movements than Dusky and other *Phylloscopus* warblers. **Status** Rare passage migrant in northern and eastern Mongolia, late April to mid-September.

adult

Dusky Warbler

adult

Sulphur-bellied Warbler

adult

Yellow-streaked Warbler

breeding

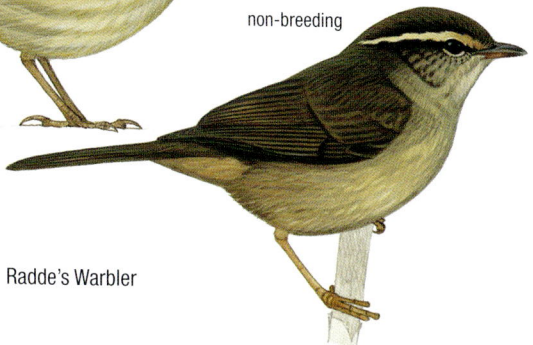

non-breeding

Radde's Warbler

Willow Warbler *Phylloscopus trochilus*

11–12 cm

ID Rather plain, without wing-bars or strong head pattern aside from whitish supercilium. Adult is whitish below with tinge of yellow on breast. Imm has overall yellowish cast. Legs pale brown. Compare Wood Warbler and Chiffchaff with which it is most likely to be confused. **Voice** Song begins with sharp, fast notes and descends into sweet warble. Contact call is a two-syllable *hooeet*. **Habitat** Deciduous and mixed forest with thick undergrowth, as well as willow copses and wooded parks and gardens. **Behaviour** Typically forages high in tree-tops, though nests on or near ground. **Status** Rare and local breeding visitor and passage migrant in the far north, in western Hövsgöl mountain range and Darkhad Depression; may also nest in Tes River valley (northern Uvs Depression). Occurs late April to early September. **Taxonomy** Two races occur: western *P. t. acredula* (vagrant) and eastern *P. t. yakutensis*.

Common Chiffchaff *Phylloscopus collybita*

10–12 cm

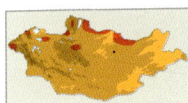

ID Generally plain though sometimes shows a single indistinct narrow wing-bar. Quite similar to Willow Warbler, but grey-brown (vs. greenish) above with buffy (vs. yellowish) tinge on cheeks, neck and sides. Legs and most of bill black (vs. pale brown). **Voice** Song is a distinctive, unmusical, repeated *chiff-chaff-chaff chiff-chiff-chaff-chaff*..., more rapid than that of European race. Usual *hweet* call is a shorter and less 'questioning' version of Willow Warbler's. **Habitat** Breeds in coniferous, deciduous and mixed forest. Migrants occur in wooded parks and gardens as well as scrub habitats. **Behaviour** Constantly bobs tail, unlike Willow Warbler. Builds domed nest on ground. **Status** Uncommon to rare breeding visitor in all northern and western mountain ranges and uncommon passage migrant throughout the country, late April to mid-September. **Taxonomy** The subspecies that occurs in Mongolia is *P. c. tristis*, Siberian Chiffchaff, but *P. c. abietinus* might occur as a rare migrant or vagrant.

Wood Warbler *Phylloscopus sibilatrix*

11–12.5 cm

ID Like colourful version of Willow Warbler with bright yellow face, throat and upper breast and contrasting clear white underparts. Imm is duller. **Voice** Song is a metallic trill that begins with lisping quality and descends into deeper rapid pulsing; also a descending series of its sweet, plaintive call, *pew-pew-pew-pew*... **Habitat** Breeds in deciduous and mixed forest with closed canopy and sparse understorey, but visits more open habitats on migration. **Behaviour** Forages and flycatches at all levels, especially in tree canopy. **Status** Vagrant. Three records: adult captured by ringers in deciduous forest in Delgermörön River valley, Hövsgöl Mountains, June 1994; sight record of bird in poplar tree in downtown Choibalsan, Dornod province, 11–12 September 2004; and imm in poplar plantation near Öndörkhaan town, Hentii province, 5 September 2013.

acredula

adult

yakutensis

adult

1st-winter

Willow Warbler

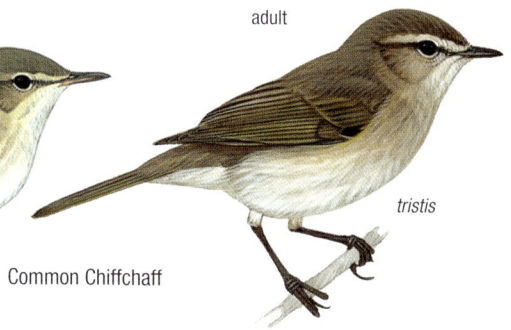

adult

adult

abietinus

tristis

Common Chiffchaff

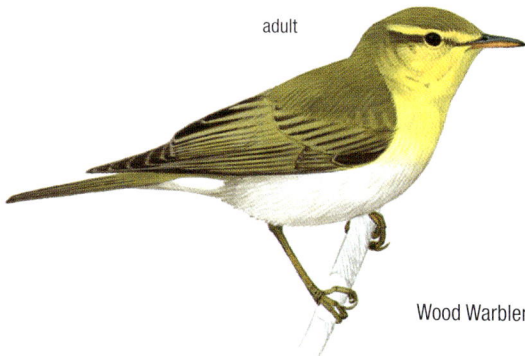

adult

Wood Warbler

Arctic Warbler *Phylloscopus borealis*

11–13 cm

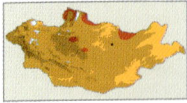

ID Easily confused with Greenish and Two-barred Warblers in general coloration and 1–2 narrow wing-bars, but has a different 'build': slightly larger, 'heavier', more elongated with long primary projection; and head notably long and flat. Whitish supercillium is thinner and much longer than in Greenish but does not extend to the forehead above the bill as in the latter. Orange lower mandible has black tip (vs. all-orange in Greenish). Compare also Pale-legged Leaf Warbler. **Voice** Song is a hard, somewhat melodious but monotone trill. Call is a sharp, buzzy *dzzik*, very different from that of Greenish. **Habitat** Typically nests in open willow-birch stands in taiga, sometimes mixed with conifers. Occurs in arid steppe and desert on migration. **Behaviour** Forages in tree-tops with little wing- or tail-flicking; nests on ground. Not shy. **Status** Uncommon to rare breeding visitor in main Hövsgöl, Khangai and Hentii mountain ranges and Orkhon-Selenge River Basin. Uncommon passage migrant throughout Mongolia, late April to early September (mid-September in the Gobi).

Greenish Warbler *Phylloscopus trochiloides*

9–10.5 cm

ID A compact, dark greenish-olive leaf warbler with 1–2 narrow wing-bars and prominent whitish supercillium. Similar to Arctic and, especially, Two-barred Warbler, which see for points of distinction. Best and safest distinctions from Arctic are build, supercillium extending to nostril, shorter primary projection and lack of diffuse streaks on breast. Siberian Chiffchaff is typically greyer (vs. greenish) above with buffy (vs. whitish) wash on face, neck and flanks, and black (vs. greyish-brown) legs. **Voice** Song is a bright jumble of liquid notes, often ending in drier, rapid trill. Call is a sweet *chs-lee*, often compared to that of White Wagtail and very different from that of Arctic Warbler and Chiffchaff. **Habitat** Thickets in deciduous and mixed forest in mountain taiga and river valleys. Occurs in steppe and desert as well as parks and gardens in town and cities on migration. **Behaviour** Constantly flicks wings and tail as it forages in tree-tops. **Status** Uncommon breeding visitor in mountain ranges across northern Mongolia, including the far east, and fairly common passage migrant throughout, late April to early September. **Taxonomy** Two-barred and Green Warblers, treated here as distinct species, were formerly considered subspecies of Greenish Warbler. Distribution and habitat of these three forms has not been clearly differentiated as yet in Mongolia.

Two-barred Warbler *Phylloscopus plumbeitarsus*

10–11 cm

ID Very similar to Greenish Warbler, from which it is probably only safely distinguished by its two distinct wing-bars. All other comparative characters are variable among individuals and under different lighting conditions and are therefore unreliable. Compare also Yellow-browed and Arctic Warblers. **Voice** Song similar to Greenish Warbler but richer; call also similar to Greenish Warbler. **Habitat** See Greenish Warbler. **Behaviour** Similar to Greenish Warbler. **Status** Status and distribution in relation to Greenish Warbler in Mongolia is poorly understood; apparently a rare and local breeding visitor to northern mountain ranges and an uncommon passage migrant elsewhere. Presumably occurs late April to early September. **Taxonomy** Considered a subspecies of Greenish Warbler *P. trochiloides* by BirdLife International and other authorities.

Green Warbler *Phylloscopus nitidus*

10–11 cm

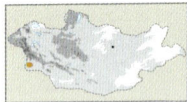

ID Distinguished from similar Greenish, Two-barred and Arctic Warblers by yellowish-green (vs. greyish-green) coloration above and yellow (vs. whitish) supercillium, cheeks and throat. The vagrant Wood Warbler is much brighter yellow on the throat and breast, bright white (vs. greyish) below, and lacks any trace of wing-bar. **Voice** Similar to Greenish Warbler, but both call and song are stronger and sweeter. **Habitat** See Greenish Warbler. **Behaviour** Similar to Greenish Warbler. **Status** Known only as a rare passage migrant (or vagrant) in SW Mongolia, but distribution and status in Mongolia are still poorly understood. Presumably occurs late April to early September. **Taxonomy** Considered a subspecies of Greenish Warbler *P. trochiloides* by BirdLife International and other authorities. [Alt: Bright Green Warbler]

1st-winter

Arctic Warbler

adult
fresh
(spring)

Greenish Warbler

1st-winter

1st-winter

Two-barred Warbler

adult

adult
fresh

Green Warbler

1st-winter

Pale-legged Leaf Warbler *Phylloscopus tenellipes* 11 cm

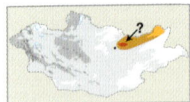

ID Resembles Arctic Warbler in large size, robust build and narrow wing-bar(s), but distinguished by pale legs, sooty-grey crown and rufescent-brown rump contrasting with greenish back. **Voice** Song is a high, thin, 'shivery', cricket-like *seeseeseeseeseesee-si-si-si*. Call is hard *tit-tit*. **Habitat** Nests in deciduous and mixed forest with heavy undergrowth and litter in mountain taiga and river valleys to 1,800 m. Uses other wooded habitats on migration. **Behaviour** Active, if somewhat sluggish forager in forest undergrowth; not shy but prefers dense cover. Often pumps tail in downward motion. **Status** Very rare local breeding visitor to Tuul and Terelj river valleys (unconfirmed breeding records). In addition, migrants have been recorded in river valleys in Hentii Mountains and Ulz River valley (Mongol Daguur steppe). Presumably present late April to late August.

Yellow-browed Warbler *Phylloscopus inornatus* 9–10.5 cm

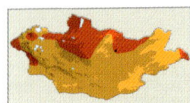

ID Similar to Pallas's Leaf Warbler but plainer, lacking that species' yellow rump and orange smudge in front of eye and showing only a faint crown-stripe. Compare also Hume's Leaf Warbler. Distinguished from all other *Phylloscopus* warblers by small size and bold double wing-bars. **Voice** Call is a high, sharp and rising *ts-weet*, repeated in song. **Habitat** Breeds in deciduous and mixed montane forest, often near water; occurs in other arboreal and scrub habitats, including wooded parks and gardens on migration. **Behaviour** Nests on ground but feeds as high as canopy. On migration, will join foraging flocks of other small woodland birds. **Status** Fairly common breeding visitor in all major mountain ranges across Mongolia except Gobi and steppe mountains, and fairly common passage migrant throughout, late April to late September. [Alt: Inornate Warbler]

Hume's Leaf Warbler *Phylloscopus humei* 9–10 cm

ID Very similar to Yellow-browed Warbler (of which it was once regarded a subspecies). It is drabber grey (vs. yellowish-green) above, with darker legs and bill and whitish (vs. yellowish) wing-bars. However, some individuals appear very similar in the field, so the two species may not be safely distinguishable by plumage characters alone. See Yellow-browed Warbler for comparison with other leaf warblers. **Voice** Call is a sweet two-note *we-soo*, very different from Yellow-browed call. This is repeated in song, which ends with long harder *tzweeeee* note. **Habitat** Breeds at higher altitude than Yellow-browed in deciduous, coniferous and mixed montane forest up to 2,700 m, typically near clearings with associated thick brush, e.g. wild rose. **Behaviour** Similar to Yellow-browed Warbler. **Status** Uncommon breeding visitor in Mongol-Altai, Hövsgöl, Khangai and Hentii mountain ranges, and common passage migrant elsewhere. Relative distribution of Hume's Leaf and Yellow-browed Warblers is poorly understood in Mongolia. However, Hume's is clearly more common in forested mountains in western than in eastern and central Mongolia. Late April to late August (to mid-September in the Gobi).

Pallas's Leaf Warbler *Phylloscopus proregulus* 9–9.5 cm

ID The most distinctive of Mongolian leaf warblers due to very small size and bold markings. The combination of contrasting crown and eye-stripes, broad double wing-bars, yellow rump and orange smudge at base of tiny black bill is diagnostic – though all of these can be hard to see from below. See similar Yellow-browed Warbler and Hume's Leaf Warbler for distinctions. **Voice** Song is a surprisingly loud, sweet, high-pitched and varied jumble of trills, twittering and some mimicry. Has been described as the 'canary of the taiga'. Call is soft and rather weak *sooee*. **Habitat** Breeds in coniferous and mixed forest in montane taiga. Occurs in other arboreal and brushy habitats, including in steppe and desert on migration. **Behaviour** Very active; often hovers while gleaning foliage and hangs upside-down like a tit. **Status** Fairly common breeding visitor in Hövsgöl, Khangai and Hentii mountain ranges, and common passage migrant throughout, late April to late September.

Goldcrest *Regulus regulus* 9–10 m

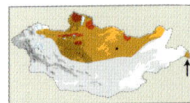

ID Smallest bird in Eurasia. Resembles *Phylloscopus* warblers, but conspicuous golden crown with black borders is diagnostic in all plumages. **Voice** Song is a brief, very hurried and shrill warbling twitter, accelerating in final flourish. Call is a very high-pitched *zi* or *zee-zee-zee*. **Habitat** Coniferous forest. **Behaviour** Forages at all levels in trees, frequently hovering and darting out to catch insects. Joins foraging flocks of other small woodland birds. **Status** Uncommon breeding visitor and passage migrant, late April to mid-October, and rare winter resident in coniferous zones of Khangai, Hövsgöl and Hentii ranges.

adult

Pale-legged Leaf Warbler

adult

adult

1st-winter

Yellow-browed Warbler

adult

1st-winter

Hume's Leaf Warbler

adult

Pallas's Leaf Warbler

♀

♂

Goldcrest

Barred Warbler *Sylvia nisoria*   15–17 cm

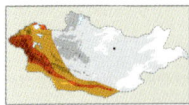

ID A distinctively large, stout, grey and heavy-billed warbler. Combination of heavily barred underparts, white wing-bars and tail-tips, and pale golden eye of adult male is diagnostic. Adult female similar but less boldly marked. Imm is browner but build, wing-bars and barred flanks and vent unlike other *Sylvia* warblers. **Voice** Song is a short, rich jumble of melodious and harsh notes sometimes given in flight. Calls are typical *Sylvia taks* or *chaks* and *chuurrs*. **Habitat** Nests in dense thickets or woodland edges with scattered bushes in high mountain valleys. **Behaviour** Fairly shy and skulking. Feeds mainly on insects but takes many berries in autumn. **Status** Uncommon and local breeding visitor at high elevations from the Mongol-Altai range south and east through the Gobi-Altai range, and uncommon passage migrant within the breeding range, late April to mid-September.

Blackcap *Sylvia atricapilla*   14 cm

ID Uniform grey coloration and black cap of male (rufous in female) highly distinctive; lacks white cheeks and black bib of superficially similar Marsh and Willow Tits. **Voice** Song is a short, variable series of energetic warbles and squeaks. Call is a crisp, toneless *tak*; also a scolding *kurrr*. **Habitat** Open woodlands, parks and gardens. **Behaviour** Not especially shy; generally forages in trees at low to mid height; sometimes sings from tree-tops. **Status** Vagrant. One recorded in mixed forest in Ögöömör valley of Terelj, Töv province, September 1997.

Common Whitethroat *Sylvia communis*   13–15 cm

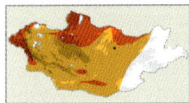

ID Adult male has distinctive combination of grey head, contrasting white throat, white eye-ring and extensive rufous in wings. Adult female and imm are duller and lack grey head. Compare Lesser and Hume's Whitethroats. **Voice** Song is a short, dry and scratchy warble, given from perch or in song flight. Call is a harsh, chiding *churr*, with more prolonged *churr-urrr-rr* given in alarm. **Habitat** Nests in open scrubby habitats at woodland and riverine edges, parks and cultivated areas. Occurs in arid steppe and desert on migration. **Behaviour** Not especially shy, readily perching and singing in open; migrating birds more skulking. **Status** Fairly common breeding visitor in most major Mongolian mountain ranges, including Gobi-Altai. Fairly common passage migrant throughout most of country, late April to mid-September.

Lesser Whitethroat *Sylvia curruca*   11–13 cm

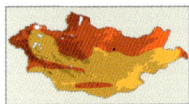

ID Smaller and greyer than Common Whitethroat with distinctive dark cheek and lacking rufous in wings of latter. Very similar to Hume's Whitethroat, which see for distinctions. **Voice** Song is a short, dry, bubbly, bunting-like babble. Call is a clicking *tek* and scolding *churr*, higher-pitched than similar calls of Common Whitethroat. **Habitat** Dense thickets and scrub at edges of forest (including mountain taiga) and lake and river margins. Occurs in arid steppe and desert on migration. **Behaviour** Shy and skulking when nesting, typically singing from deep cover; more gregarious and less retiring on migration. **Status** Fairly common breeding visitor to all major mountain ranges in Mongolia, including the far east, and fairly common passage migrant throughout, late April to early September (mid-September in the Gobi).

Hume's Whitethroat *Sylvia althaea*   11–13 cm

ID Very like Lesser Whitethroat, but slightly larger, with dark slate-grey (vs. brownish-grey) upperparts, more prominent white outer tail-feathers and heavier bill. Juv has dull plumage and iris. **Voice** Song is much richer warble – more Blackcap-like – than Lesser Whitethroat's. Calls are similar to Lesser Whitethroat's. **Habitat** Low bushes including juniper scrub on rocky slopes and gullies in high mountains. Occurs in Gobi Desert on migration. **Behaviour** Similar to Lesser Whitethroat. **Status** Very rare and local breeding visitor, so far known only from isolated records in Mongol-Altai, Gobi-Altai and Khangai mountain ranges, though possibly breeds at other high-altitude sites. Uncommon passage migrant through breeding areas and in Gobi oases, late April to early September. **Taxonomy** Sometimes considered a subspecies of Lesser Whitethroat. [Alt: Mountain Lesser Whitethroat]

Asian Desert Warbler *Sylvia nana*   11.5 cm

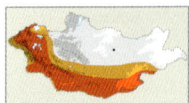

ID Combination of sandy-brown coloration with rufous rump and tail and yellow iris is unique. Behaviour and habitat are diagnostic. **Voice** Song is a short series of sweet, clear notes that has been aptly described as 'jingling' and 'slightly jerky'. Calls include a weak, falling trill and shrill *chee-chee-chee*. **Habitat** Open scrub including *Caragana* bushes in desert and rocky hills. **Behaviour** Shy, but due to open habitat, often easily seen, scuttling on ground or quickly flying between bushes. **Status** Fairly common breeding visitor and passage migrant in arid habitats from Mongol-Altai range in south and east throughout the Gobi, late April to late September. [Alt: Desert Warbler]

Barred Warbler

♂ breeding

♀ breeding

1st-winter

♂

♀

Blackcap

♂

1st-winter

Common Whitethroat

1st-winter

breeding

Lesser Whitethroat

breeding

1st-winter

Hume's Whitethroat

juvenile

breeding

Asian Desert Warbler

Reed Parrotbill *Paradoxornis heudei*   16–18 cm

ID No other Mongolian species combines this bird's disproportionately large head; white throat and cheeks; long black line from above eye to nape; heavy but stubby (parrot-like) yellow bill; and very long tail. See also habitat. **Voice** Nasal whistling and brief trilling calls. **Habitat** Reedbeds. **Behaviour** Agile and tit-like. Forages for seeds in small, active flocks. **Status** Rare and very local resident breeder, known only from Lake Tashgain Tavan, Lake Buir and Khalkh River delta, Dornod province. **Conservation** Considered Near Threatened globally and Endangered in Mongolia due to restricted range, destruction of reedbed habitat and consequent population decline. Listed in the Threatened Birds of Asia (2001) and Mongolian Red Book (2013).

Vinous-throated Parrotbill *Sinosuthora webbianus*   12–13 cm

ID Might at first be mistaken for warbler or finch, but combination of tiny, conical black bill, plain face, 'wine-coloured' plumage and long tail are diagnostic. See also behaviour. **Voice** Song is a repeated high-pitched *reit-reit pevii-u vee-ei-ei*. Contact call is a high *chi-chi-chi*. **Habitat** High grass reedbeds, shrub thickets and forest edges. **Behaviour** Forages very actively for insects, fruits and seeds in constantly chattering flocks. **Status** Vagrant. One collected, Khalkh River and Lake Tashgai, Dornod province, 20 March 1970. Uncertain records from Khalkh, Nömrög, Degee and Altan River basins, Dornod province. **Taxonomy** Formerly placed in genus *Suthora* or *Paradoxornis*.

Bearded Reedling *Panurus biarmicus*   14–17 cm

ID Adult male unmistakable with powder-blue head, accented with black moustache and undertail-coverts; thin yellow bill; golden iris; general yellowish-rufous plumage above; long, pointed tail; and white wing-patches. Juv and adult female lack blue head and moustache. Juv has black back and sides of tail and darker iris. **Voice** Song is a series of short, trisyllabic and rather toneless notes. Contact call a hard metallic *ping*. **Habitat** Reedbeds. **Behaviour** Moves rather unobtrusively through reeds, typically in family groups and post-breeding flocks. Feeds on reed seeds and insects. **Status** Uncommon and local resident breeder in scattered localities across Mongolia, including Great Lakes Depression, Valley of the Lakes, Orkhon-Selenge River Basin, Khangai and Hentii mountain range, Buir Lake-Khalkh River-Khyangan Region and possibly South Shargyn and Trans-Altai Gobi. [Alt: Bearded Parrotbill, Bearded Tit]

adult

polivanovi

Reed Parrotbill

adult

Vinous-throated Parrotbill

♂ juvenile

♀

♂

Bearded Reedling

Long-tailed Tit *Aegithalos caudatus*   14–16 cm

ID Unmistakable: tiny pinkish-and-white ball of fluff with very small, black bill and absurdly long tail. **Voice** Song is a soft, thin twittering, rarely heard. Contact calls are variations on soft *tip* and buzzy *trrrt*. **Habitat** Coniferous and mixed forest. **Behaviour** Forages in family groups when not nesting. May flock with other tits. Agile, lively and restless as it moves through thickets. In winter, descends from highland taiga to forested steppe, river valleys and town centres. **Status** Uncommon resident breeder, mainly in north-central Mongolia, but with disjunct populations in southwest (Mongol-Altai range) and far eastern Dornod.

White-crowned Penduline Tit *Remiz coronatus*   11–15 cm

ID Adult male is distinctive, with whitish head set off by black mask and head-band in combination with bright rufous back and very fine, dark bill. Adult female is similar, but duller with less contrasting black mask. Juv is similar to adult female but duller and lacks black mask – a rather anonymous little bird – but fine bill, rusty bend of wing, habitat and behaviour should clinch its identity. **Voice** A thin, plaintive *tseeeooo* is commonest call; also gives *tslu-su-su* and buzzing *tzzz*. Song combines several call types interspersed with trills. **Habitat** Willow and poplar groves along rivers and other wetland borders. On migration, visits wooded parks and gardens. **Behaviour** Feeds actively in tops of trees and shrubs, in tit-like manner. Nest is remarkable, pendulous, flask-shaped construction of strong felt-like material made from wool of willow and poplar seeds reinforced with plant fibres. **Status** Fairly common but localised breeding visitor and uncommon passage migrant in wooded river valleys across Mongolia, late April to late August (early September in the Gobi). **Taxonomy** Formerly considered a subspecies of Eurasian Penduline Tit *R. pendulinus*.

Eurasian Nuthatch *Sitta europaea*   12–14 cm

ID Mongolia's only nuthatch and therefore instantly recognisable by its unique, 'neckless' shape, and foraging behaviour – often perching head downward on tree trunks – as well as its blue-grey upperparts and rusty lower flanks and undertail-coverts. **Voice** Song is a steady whistling *pwee-pwee-pwee-pwee*, sometimes accelerating to *pipipipipipipipi*. Call is a ringing *tuit tuit*, also gives a softer *sit*. **Habitat** Coniferous, deciduous and mixed forest, as well as wooded parks and gardens, especially in winter. **Behaviour** Actively forages over tree trunks and branches with great agility; unlike woodpeckers and treecreepers can descend head-first. Feeds on insects and also wedges nuts into bark cracks and hacks them open. **Status** Fairly common resident breeder across northern (forested) half of Mongolia. [Alt: Wood Nuthatch]

Eurasian Treecreeper *Certhia familiaris*   12–15 cm

ID Combination of small size; cryptic brown-variegated upperparts and clear white underparts; long, decurved bill; long, stiff, pointed tail; and creeping behaviour make it unlike any other Mongolian species. **Voice** Song is a very high, but descending trilling warble. Call is also a very high-pitched, thin Goldcrest-like *tseee*. **Habitat** Coniferous and mixed forest. **Behaviour** Climbs up tree trunks in spiral, probing cracks in bark for insects. Flies from height to base of next tree. **Status** Uncommon resident breeder in main forested ranges throughout.

Wallcreeper *Tichodroma muraria*   15–17 cm

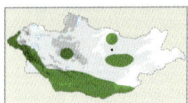

ID Unique and unmistakable in form, plumage pattern, and behaviour, its bright red-and-grey polka-dotted 'butterfly' wings distinguishing it from any other bird. In breeding plumage, adult shows black throat and upper breast, but in non-breeding plumage these areas are white. **Voice** Song is a slowly delivered combination of breathy whistles. Various calls, including piping *tui*, soft *chirp* and buzzing *zree*. **Habitat** Sheer cliff faces in high mountains near mountain water courses. **Behaviour** Scales rock faces, actively probing crevices for invertebrates, frequently flicking wings; flutters and glides in brief flights that show off its spectacular wings. **Status** Uncommon to rare and very local resident breeder in Mongol-Altai, Gobi-Altai, Khangai and Hentii ranges. Descends from high-altitude breeding areas to valleys and rocky lakeshore slopes and rarely mountain ranges in steppe in winter.

adult

Long-tailed Tit

juvenile

White-crowned Penduline Tit

1st-winter

♂
variant

♂

adult

Eurasian Nuthatch

♀/
non-breeding

♂
breeding

Wallcreeper

adult

Eurasian Treecreeper

Pale Thrush *Turdus pallidus*

22–24 cm

ID In Mongolia, this distinctively grey-headed, rufous-sided thrush is likely to be confused only with Eyebrowed Thrush, but lacks latter's prominent white supercilium. Adult male has darker, more contrasting coloration. **Voice** Song is a three-note repetitive whistle, interspersed with brief chitters. Also gives thin *tzee* and stronger *chrrri-ii-ip* calls, and low *ko-ko-ko*. **Habitat** Nests in montane pine and deciduous forest with thickets and secondary growth. Occurs in other habitats on migration. **Behaviour** Feeds in fruiting trees as well as on ground. **Status** Very rare breeding visitor in Hentii and Khangai mountain ranges, and rare passage migrant through breeding range, presumably late April to mid-September.

Eyebrowed Thrush *Turdus obscurus*

21–23 cm

ID Similar in general coloration to Pale Thrush but with diagnostic white supercilium. Sexes similar but adult male has deeper colour. 1st-winter is similar to adult female but has pale wing-bar and is duller. **Voice** Song is a simple, repeated, trisyllabic whistled phrase consisting of rather abrupt notes. Also makes high thin *seep* contact calls, and a briefer *sip-sip*. **Habitat** Nests in moist montane spruce and birch forest and willow thickets near rivers and streams. Migrants also occur in steppe and desert. **Behaviour** Rather shy. Feeds on ground but quickly flies into cover if disturbed. **Status** Scarce breeding visitor in Hövsgöl, Khangai and Hentii ranges, and uncommon passage migrant countrywide, late April to late September.

Siberian Thrush *Geokichla sibirica*

20–23 cm

ID Overall slate-grey plumage of adult male, with contrasting white supercilium, is unmistakable. 1st-winter male similar to adult male but duller with paler markings. Adult female is plain olive-brown above with heavy black scaling below, creating possible confusion with White's Thrush. However, the latter is notably larger with longer bill and tail and is densely scaly above as well as below. Scaly female rock thrushes are much smaller and more compact, and occur in open (vs. forested) habitats. **Voice** Song has been described as 'rich, whistled phrases...*tve-kwi-tring* or similar'. Calls include a low clucking when flushed, a high-pitched *tsee* and a quiet *zit*. **Habitat** Breeds in dense coniferous and mixed forest with dense understorey shrubs. Occurs in steppe and arid mountains on migration. **Behaviour** Rather shy and skulking, but may flock with other thrushes on migration. **Status** Very rare and local breeding visitor, known to nest only in northernmost Hövsgöl and Selenge provinces, and rare passage migrant through all major mountain ranges, late April to early September. **Taxonomy** Formerly placed in genus *Zoothera*.

Pale Thrush

Eyebrowed Thrush

♂

1st-year

♀

1st-year

♂

Siberian Thrush

♂

Black-throated Thrush *Turdus atrogularis*

24–26 cm

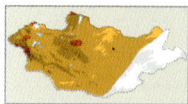

ID Adult male is uniform medium grey above and white (obscurely streaked) below, with black face, throat and upper breast. In flight, shows rufous underwing-coverts. Adult female is less well marked on head and breast. 1st-winter birds have distinct streaking on breast and flanks. **Voice** Song is a series of short, fluty, rambling phrases, *churee-weeoo*, etc., interspersed with squeaky notes, said to be harsher than that of Red-throated, with more frequent pauses. Typical *Turdus* calls include a harsh *chack* and thin *seep*. **Habitat** Breeds in mixed forest in taiga and shrubby montane woodlands. Occurs in steppe and desert on migration. **Behaviour** Builds mud-lined nest in trees or on ground. Wintering birds survive on fruits such as Sea Buckthorn. **Status** Uncommon to rare, with restricted breeding range from Great Lakes Depression and Mongol-Altai range east to Hövsgöl and Khangai ranges. Migrates throughout rest of the country and, depending on temperature and snow cover, may remain in winter, especially in parks and gardens in towns and cities. Present late April to late September and sometimes in winter. **Taxonomy** Formerly considered conspecific with Red-throated Thrush.

Red-throated Thrush *Turdus ruficollis*

24–26 cm

ID Adult male is uniform medium grey above and white (obscurely streaked) below, with bright rufous face, throat and upper breast. In flight, shows rufous underwing-coverts. Also differs from Black-throated in having rufous tail. Adult female is less colourful on head and breast. 1st-winter birds have distinct streaking on breast and flanks. **Voice** Song is a series of short, fluty, rambling phrases, *churee-weeoo*, etc., interspersed with squeaky notes. Typical *Turdus* calls include a harsh *chack* and thin *seep*. **Habitat** Breeds in mixed forest in taiga and shrubby montane woodlands. Occurs in steppe and desert on migration. **Behaviour** Builds mud-lined nest in trees or on ground. Wintering birds survive on fruits such as Sea Buckthorn. **Status** Fairly common breeding visitor in all major mountain ranges across northern Mongolia. Migrates throughout rest of the country and, depending on temperature and snow cover, may remain in winter, especially in parks and gardens in towns and cities. Present late April to late September and sometimes in winter. **Taxonomy** Formerly considered conspecific with Black-throated Thrush.

Dusky Thrush *Turdus eunomus*

23–25 cm

ID Complex, variegated plumage, unlikely to be confused with other Mongolian thrushes, boldly patterned in black and white on head and underparts (replacing rufous markings of Naumann's Thrush), with extensive reddish-brown on upperwings. **Voice** Song is a mellow, clear, descending warble, with a twittering flourish at the end. Calls include a harsh *chak* and Common Starling-like *skiiiir*. **Habitat** Breeds in mixed open taiga forest with some combination of larch, pine, willow, poplar and birch; also occurs in steppe and desert on migration. **Behaviour** Shy and wary, feeding mainly on ground and in fruiting trees. Flocks with other thrush species during migration. **Status** Generally nests north of Naumann's Thrush as far as Siberian tundra line; it may nest in montane forests of Khangai and Hentii ranges, but there are no documented breeding records. Present late April to mid-September. **Taxonomy** Formerly considered conspecific with Naumann's Thrush.

Naumann's Thrush *Turdus naumanni*

23–25 cm

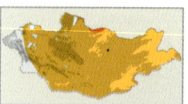

ID Complex, variegated plumage pattern, with heavy bright rufous spotting below, rufous tail and grey-brown upperparts. Dusky Thrush is boldly patterned in black and white on head and underparts, this essentially replacing the rufous markings of Naumann's. Intermediate plumages occur in broad zone where the two species overlap. **Voice** Song is a mellow, clear, descending warble, with a twittering flourish at the end. Calls include a harsh *chak* and Common Starling-like *skiiiir*. **Habitat** Breeds in mixed open taiga forest with some combination of larch, pine, willow, poplar and birch; also occurs in steppe and desert on migration. **Behaviour** Shy and wary, feeding mainly on ground and in fruiting trees. Flocks with other thrush species during migration. **Status** More southerly distribution than Dusky Thrush, nesting mainly in southern Siberia. Rare and local breeding visitor to Zelter River valley, Selenge province, and uncommon migrant throughout rest of country. Present late April to mid-September. **Taxonomy** Formerly considered conspecific with Dusky Thrush.

Fieldfare *Turdus pilaris*

25–27 cm

ID No other Mongolian thrush combines largely grey upperparts with contrasting dark brown back and blackish tail and heavily streaked and spotted underparts with tawny wash on breast. Sexes and ages similar. 1st-winter is duller, with paler orange breast. **Voice** Song is an unmelodious dry warbling with squeaking and chuckling phrases. Calls include a soft, peevish clucking chuckle, *chuck-chuck-chuck*; also a harsh, rattling alarm call. **Habitat** Nests at margins of a variety of forest types in mountains and river valleys. Inhabits brushy steppe and desert as well as parks and gardens with fruiting trees at lower elevations outside breeding season. **Behaviour** Feeds readily in open; not especially shy. **Status** Uncommon breeding visitor in northern Uvs Depression and Orkhon-Selenge River Basin. On migration, occurs widely in areas with fruiting trees including Gobi-Altai range. Uncommon to rare winter visitor, depending on severity of weather, in central and western Mongolia, often entering cities and towns to feed on ornamental fruit trees. Migrant population present late April to late September.

♂

1st-winter

♂

Black-throated Thrush

♂

♂

1st-winter

Red-throated Thrush

♂

1st-winter

♂

Dusky Thrush

♂

1st-winter

♂

♂

adult

Naumann's Thrush

Fieldfare

Common Blackbird *Turdus merula*   25–28 cm

ID Adult male is only Mongolian thrush that is all-black with yellow bill. However, compare adult breeding Common Starling. Adult female and imm are dark brown with darker streaking and lack yellow bill, but overall dark coloration remains distinctive. **Voice** Song consists of very rich and mellow unhurried warbling phrases. Call is a low *chuk-chuk-chuk*, accelerating and rising in pitch and volume in alarm to an explosive *chik-chik-chik-chink-chink*. Also gives a thin, breathy *seee* in flight. **Habitat** Open forest and forest edges, as well as parks and garden in cities and towns. **Behaviour** Forages mainly on ground, feeding on small vertebrates as well as insects and fruits. Moves on ground in rapid bounds. **Status** Vagrant. Known from only two records: a partial carcass found in Davaany Zörlög, 10 km west of Ulaanbaatar, 10 September 1988 and a female photographed in Khovd city, Khovd province, 9 May 2007. [Alt: Eurasian Blackbird]

Redwing *Turdus iliacus*   23–25 cm

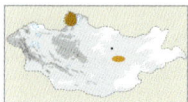

ID In all plumages, combination of broad white supercilium, dark streaking (vs. spotting) below, and dark rufous flanks is unique among Mongolian thrushes. **Voice** Song is highly variable, but typically begins with a few loud fluty notes and ends with high-pitched twittering. Flight call is a thin buzzy *tzeee*; alarm call a rattling *chittick*. **Habitat** Coniferous forest. Occurs in steppe on migration. **Behaviour** Shy and skulking when breeding. Gregarious and more likely to be seen in open steppe on migration, often in company of other thrushes. **Status** Apparently very rare breeding visitor in Darkhad and northern Hövsgöl mountain range, but nesting as yet unconfirmed. Probably a rare passage migrant in Bayantsagaan district, Töv province, and Choir district, Gobisumber province. No records exist outside breeding areas.

Song Thrush *Turdus philomelos*   22–25 cm

ID One of only two Mongolian thrushes with an evenly spotted (vs. streaked or scaly) breast. Distinguished from similarly spotted Mistle Thrush by its much smaller size; brownish (vs. greyish) upperparts; buffy wash on upper breast; and 'arrowhead' (vs. Mistle's rounded) spots. In flight, distinguished from Redwing by rusty-buff underwing. **Voice** Song is a series of short, emphatic warbles and whistles, each repeated twice or more and interspersed with harsher sounds. Flight call is a barely noticeable *zip*; alarm call a louder *chick*. **Habitat** Nests at margins of coniferous and mixed taiga forest and in thick shrubbery; associated with fruiting trees and shrubs at lower elevation and in steppe post-breeding. **Behaviour** Feeds on ground. Takes all manner of invertebrates but is specialist predator of snails, striking them against a favourite stone ('thrush's anvil') to break their shells. **Status** Very rare breeding visitor in valleys of Tes and Torkhilog Rivers in Northern Uvs Depression. Very rare passage migrant through Great Lakes Depression and southern Hövsgöl and Khangai ranges; presumably present late April to early September.

Mistle Thrush *Turdus viscivorus*   24–26 cm

ID Likely to be mistaken only for smaller, browner Song Thrush, which see for further distinctions. Note white corners on tip of tail in flight. **Voice** Song is a mellow, fluty whistle with burred notes similar to that of Common Blackbird, but more monotonous. Call is a hard rattling chatter given in flight and alarm. **Habitat** Open montane woodland with scattered large trees; occurs in steppe and desert on migration. **Behaviour** Often feeds in open, though rather shy. Aggressively territorial when breeding; somewhat gregarious when not. **Status** Very rare and local breeding visitor in river valleys of the Mongol-Altai mountain range. Recorded as rare passage migrant in Darkhad Depression, Hövsgöl province, Khangai mountain range and Bulgan River valley in Baruunkhurai Depression, as well as Mongol-Altai range; presumably late April to early September.

White's Thrush *Zoothera aurea*   27–31 cm

ID Large size, with notably long tail and bill and 'pot-bellied' look, together with heavy black scaling above and below, make confusion with other species unlikely. However, see Siberian Thrush for distinction from other scaly species. Sexes similar. **Voice** Distinctive song consists of haunting, melancholic, widely spaced, single-note whistles that fade away gradually. Call is a thin, whistled *tsi*. **Habitat** Nests in dense montane coniferous and mixed forest with thick, shrubby understorey. Occurs in rocky and bushy steppe and desert on migration. **Behaviour** Feeds on ground, walking with distinctive 'sneaking' stride rather than hopping. Generally shy, preferring to keep to dense cover, though typically sings from prominent high perches. **Status** Rare and local breeding visitor, known to nest only in extreme north at Torkholig River (northern Uvs Depression) and north of Lake Hövsgöl. Uncommon passage migrant throughout the rest of Mongolia, late April to mid-September (late September in the Gobi). **Taxonomy** Formerly treated as a subspecies of extralimital Scaly Thrush *Z. dauma*.

♂

♂

adult

♀

adult

Common Blackbird

adult

adult

Redwing

Song Thrush

juvenile

adult

Mistle Thrush

adult

1st-winter

adult

adult

White's Thrush

Common Redstart *Phoenicurus phoenicurus*   12–13 cm

ID Adult male distinguished from all other chat-like birds by combination of black throat; bright rufous breast and belly; bluish-grey crown and nape; and rufous tail which it 'shivers' conspicuously. Rufous tail also separates otherwise plain female and imm from other chats. For differences from other redstarts, see those species. **Voice** Song is a variable, sweet, high-pitched warbling. Call is a soft *hooeet*, similar to that of Willow Warbler; sometimes *hooeet-tek-tek* when alarmed. **Habitat** Nests in mixed open montane forest and forest edge and thickly vegetated river valleys. Occurs in shrubbery in steppe and desert on migration. **Behaviour** Places cup nest in the hollow of tree, or on ledge of building. Flycatches and gleans leaves, in forest canopy, middle levels and on ground. **Status** Uncommon breeding visitor in forested mountains from Great Lakes Depression and Mongol-Altai to Hövsgöl, Khangai and Hentii ranges, and uncommon passage migrant throughout, late April to late August (early September in the steppe).

Black Redstart *Phoenicurus ochruros*   12–14 cm

ID Adult male distinguished from other Mongolian redstarts by black breast (as well as throat) and darker mantle; lacks white wing markings. Adult female and imm are dark sooty-grey with brown tinge. **Voice** Song is a brief, hurried high warble, followed by a bizarre grating sound. Assorted brief calls include *tuc* and *tsip*. **Habitat** Cliffs and rocks with scattered shrubs in high mountain ranges at 3,000–3,200 m in both dry and relatively moist regions. Occurs at lower altitudes and at desert oases on migration. **Behaviour** Nests in rocky mountains near water sources. Builds a cup nest in hollow of rock face. Feeds mainly on ground, hopping rapidly and pausing in upright stance. Shivers tail frequently. **Status** Uncommon breeding visitor to high elevations in Mongol-Altai, Gobi-Altai, Hövsgöl, Khangai and Hentii ranges. Scarce migrant throughout, including eastern Mongolia, late April to late August (early September in the Gobi). The race in Mongolia is *P. o. phoenicuroides*.

Eversmann's Redstart *Phoenicurus erythronotus*   13–14 cm

ID Adult male distinguished from other Mongolian redstarts by unique combination of rufous throat and back; and both a white 'patch' at base of primaries and white 'slash' across the wing-coverts. Adult female and imm distinguished by buffy double wing-bars. **Voice** Song is sweet, high, rapid warble, *few-eet*, largely lacking the harsher sounds of other redstart songs. Calls are a soft *tsip* and harder and more nasal *tcharr*. **Habitat** Moist coniferous and mixed forest with thick undergrowth along wet cliffs or wet hillsides of subalpine zone at 1,800–2,400 m. **Behaviour** Places cup nest in hollows in banks, or dead snags, under tree roots or in rock crevices. Perches prominently on trees and tall bushes and drops to ground to feed. Jerks rather than shivers its tail. **Status** Rather rare and local breeding visitor to scattered localities in Mongol-Altai, Hövsgöl and Khangai ranges, and scarce passage migrant elsewhere, late April to late August (early September in the south). [Alt: Rufous-backed Redstart]

Daurian Redstart *Phoenicurus auroreus*   13–14 cm

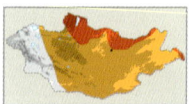

ID All plumages similar to Common Redstart, but easily distinguished by conspicuous triangular white wing-patch, which shows as white band in flight, and black (vs. grey) back. The white patch is unique among adult female and imm redstarts in Mongolia. Adult male Güldenstädt's Redstart is significantly larger, has white cap and is rarely found in same habitat. Adult male Eversmann's Redstart has rufous (vs. black) throat and back and different pattern of white wing markings. **Voice** Song is a high, cheery warble interspersed with tweets and scratchy sounds. Calls are either high and sharp or hard, e.g. *weet* and *tak-tak*. **Habitat** Mature coniferous and mixed forest in well-watered areas in a fairly wide altitude range. Also occurs in steppe and gardens and parks on migration. **Behaviour** Makes cup nest in hollow of tree, rock face or artificial structure in towns and villages. Mainly insectivorous, feeding on ground as well as in trees. Frequently shivers its tail. **Status** Locally common breeding visitor to mid-elevation mountains in Hövsgöl, Khangai and Hentii ranges, as well as locally in eastern Mongolia (Dornod province). Common passage migrant throughout, late April to late September.

♂ fresh (autumn)

♀

♂

♂

Common Redstart

Black Redstart

♀

♂

phoenicuroides

♂ fresh (autumn)

♀

♂ worn (summer)

Eversmann's Redstart

♀

♂ 1st-winter

♂

♂

Daurian Redstart

Güldenstädt's Redstart *Phoenicurus erythrogastrus*   16–18 cm

ID A very large alpine redstart. Unique combination of white cap, black back and broad white wing-patch make adult male unmistakable. Adult female and imm are similar to Black Redstart but in addition to large size, are distinguished by paler coloration with brownish (vs. sooty-grey) tone, and trace of male's wing-patch. **Voice** Song is a short series of clear, sweet, chirpy warbles with little harshness. Calls are a loud and hard *tac tac tac* or *tyek tek tek*. **Habitat** Rocky slopes in high alpine zone at 3,000 m or more, up to edge of snow-line, descending to lower valleys, foothills, steppe and desert during unfavourable weather and in winter. **Behaviour** Nest is bulky cup sited on rocky ledge or on ground. Post breeding, small flocks gather at lower altitudes to feed on winter fruits, e.g. of Sea Buckthorn. **Status** Uncommon and local resident breeder, passage migrant and winter visitor, nesting in alpine zone in Mongol-Altai, Gobi-Altai, Hövsgöl, Khangai and Hentii ranges. [Alt: White-winged Redstart]

European Robin *Erithacus rubecula*   14 cm

ID A distinctive, rotund little chat with proportionally large head and bright orange-rufous face, throat and breast; unlike any other Mongolian species. Juv is brown with paler speckling on head and back, scaly breast and brown tail. **Voice** Song is a sweet, melancholic twittering, with slow and fast phases. Typical call is a hard *tic*, or *tic-tic-tic*. Alarm call is a thin, high *seeee*. **Habitat** Typically, a bird of parks and gardens as well as open woodlands. **Behaviour** Forages on ground. Often especially active at first and last light. Not shy. **Status** Vagrant. Two different birds were photographed at Otson Chuluu, Khovd province, 15–17 November 2007.

Common Nightingale *Luscinia megarhynchos*   14–15 cm

ID Best distinguished from other thrush-like birds in Mongolia by its notably plain (unmarked) plumage: brown above with bright rufous rump and tail; and whitish below with buffy wash on breast. In flight, tail colour might suggest female redstart, but Common Nightingale is larger, with richer coloration and pinkish-brown (vs. blackish) legs. **Voice** The famous song, typically sung from dense cover, consists of remarkably rich, powerful and mellow short phrases, including rapid warbles, a slow, low-pitched *jug-jug-jug* and thin high *tseeee-tseeee-tseeee*. Calls include a low growling *trrrr*, or *seeee* in alarm. **Habitat** Deciduous woodland with dense, shrubby undergrowth. **Behaviour** Shy and skulking. Nests on or just off ground. Feeds in leaf litter as well as in low vegetation. **Status** Rare and very local breeding visitor and passage migrant; known only from Bulgan River valley, Baruunkhurai Depression, Khovd province, late April to late August. [Alt: Rufous Nightingale]

Siberian Rubythroat *Calliope calliope*   14–16 cm

ID Bright, shining red throat with thin black border and white supercilium and moustachial stripe make adult male unmistakable. Adult female and imm have plain, uniform dull-brown plumage with a more subdued head pattern; a few females have trace of red throat. Distinguished from female redstarts by whitish supercilium and brown (vs. rufous) tail, and from Bluethroat by clear (vs. boldly marked) throat and breast pattern. **Voice** Song is a rich, melodious series of warbles and whistles, rather like Common Nightingale but 'thinner', less powerful. Call is a low clucking *chak chak*. **Habitat** Nests among bushes and thick undergrowth, in or near deciduous and mixed forest, often near water. On migration, occurs in steppe and desert. **Behaviour** Shy while nesting, singing from deep cover; much less so on migration. Often cocks its tail. **Status** Fairly common breeding visitor in northern and eastern mountains from northern Great Lakes Depression east through Hövsgöl, Khangai and Hentii ranges to Buir Lake-Khalkh River-Khyangan Region in far east and sporadically common passage migrant throughout the country, late April to late September. **Taxonomy** Formerly placed in genus *Luscinia*.

♂

♀

♂

Güldenstädt's Redstart

adult

European Robin

adult

Common Nightingale

♀
variant

♀
1st-winter

♂

♂

Siberian Rubythroat

Bluethroat *Luscinia svecica*   14–15 cm

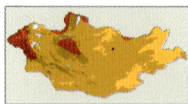

ID Bright blue, rufous and black throat and breast of adult male are diagnostic, as is the black-and-rufous tail pattern in flight. Overall brownish coloration and rufous tail-base of imm (and many females) may suggest female redstarts, but the latter lack Bluethroat's strong head and throat markings. **Voice** Song is a highly variable combination of sweet whistles and chattering, including mimicry of other birds; often incorporates a bell-like *ting ting ting*. Call is a hard *tchak tchack*, also a softer plaintive *hweet*. **Habitat** Nests and forages in willow and other thickets near conifer forest in subalpine zone of high mountains; often near streams and in vegetated marshy areas close to large lakes. Occurs in steppe and desert during migration. **Behaviour** Nests and feeds mainly on ground or in low vegetation and thickets. Typically holds its tail cocked and wings drooping. Often sings from top of bush. **Status** Uncommon and local breeding visitor from Great Lakes Depression and Mongol-Altai range through Hövsgöl, Khangai and Hentii ranges and uncommon passage migrant throughout, late April to mid-September. **Taxonomy** Three subspecies, *L. s. pallidogularis*, *L. s. saturatior* and *L. s. kobdensis* occur in Mongolia.

Siberian Blue Robin *Larvivora cyane*   13–14 cm

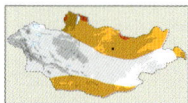

ID Combination of cobalt-blue upperparts with black border and snowy-white underparts of adult male is unique among Mongolian chats and thrushes. Imm male always shows some trace of blue on mantle and tail. Adult female perhaps best distinguished by its plainness (absence of diagnostic plumage characters) but usually shows some blue on rump and tail. Adult female Siberian Rubythroat is similar but has contrasting pale throat and supercilium; Common Nightingale has rufous tail etc. **Voice** Song begins with hesitant *tsit tsit tsit* before breaking into explosive, sweet staccato warble. Call is a crisp soft *chak*. **Habitat** Nests in coniferous and mixed montane taiga forest. Occurs in brushy steppe and desert as well as plantations in cities and towns on migration. **Behaviour** Very shy and skulking on breeding grounds; much less so where cover is scant during migration. Forages mainly on ground but climbs into higher branches to sing. **Status** Rare breeding visitor in Hövsgöl and Hentii ranges; rare passage migrant elsewhere, including Gobi and far east, late April to early September. **Taxonomy** Formerly placed in genus *Luscinia*.

Rufous-tailed Robin *Larvivora sibilans*   13 cm

ID Combination of bright rufous tail and in wings above, and scaly breast and flanks distinguish this accidental species from Common Nightingale and other brownish thrush-like birds in Mongolia. **Voice** Song is a long descending trill, somewhat reminiscent of Little Grebe. Call is a brief *chirrup*. **Habitat** Deciduous and conifer forest with thickets. **Behaviour** Skulking, foraging mainly on ground. **Status** Vagrant. Two records: one in Khalkh River valley, Dornod province, May 1987 and one at Sharyn Gol in the Selenge River basin, Selenge province (undated). **Taxonomy** Formerly placed in genus *Luscinia*. [Alt: Swinhoe's Robin]

Red-flanked Bluetail *Tarsiger cyanurus*   13–14 cm

ID Adult male readily distinguished from other small Mongolian robins and thrushes by combination of blue upperparts and orange flanks. Adult female and 1st-winter birds are much duller than adult male but retain bright blue tail (though can be hard to see colour in typical dark forest habitat) and orange flanks, as well as prominent white eye-ring and throat. **Voice** Song is a short, rich, rapid warble. Calls include a sharp *tac-tac* and soft *huit*. **Habitat** Nests in dense, moist coniferous forest and birch woods with thick undergrowth. Migrates through brushy steppe and desert. **Behaviour** Nests on or near ground in natural hollows. Not as skulking as Siberian Rubythroat or Common Nightingale, but stays in forest shade when nesting. Often sings from prominent tree-top perch, however. Feeds on the ground and also gleans and flycatches. **Status** Common breeding visitor to northern Mongolia in Hövsgöl, Khangai, Buteel and Hentii mountain ranges and very common passage migrant throughout the country, mid-April to late September. [Alt: Orange-flanked Bush Robin]

♂
non-breeding

♀

♂
breeding

pallidogularis

Bluethroat

♀
1st-winter

♀

Siberian Blue Robin

♂

adult

♀

♂

Rufous-tailed Robin

Red-flanked Bluetail

Siberian Stonechat *Saxicola maurus* 12–13 cm

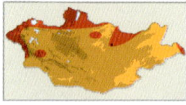

ID With one exception, the bold, contrasting black, white and orange plumage of adult male is unmistakable. See very similar White-throated Bush Chat for distinctions. Adult female is much duller and streaked brown above, but retains a distinguishing dark tail, white wing-patch and buffy breast. 1st-winter is very similar to adult female but duller and with buffy back. **Voice** Song is a brief high, squeaky warble. Usual call is a hard, stone-tapping *chak*, often accompanied by whistle: *weet-chak-chak*. **Habitat** Breeds in wet, open areas with low scattered shrubs in valleys of taiga forest and high mountains. Occurs in steppe and desert on migration. **Behaviour** Perches conspicuously on tops of bushes, prominent rocks, etc. Frequently flicks its wings. **Status** Uncommon breeding visitor in suitable habitats in high mountain ranges and uncommon to fairly common passage migrant throughout country, late April to mid-September. **Taxonomy** Considered to be a subspecies of Common Stonechat *S. torquata* by BirdLife International and other authorities. IOC considers *S. m. stejnegeri* to be a separate species (Stejneger's Stonechat) but many birds are indistinguishable from *S. maurus*.

White-throated Bush Chat *Saxicola insignis* 14.5 cm

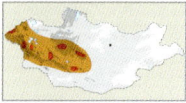

ID Very similar to Siberian Stonechat, but larger. Adult male has white (vs. black) throat and white stripe in primaries as well as inner wing, and more extensive orange on breast. Adult female is earth grey-brown, with distinct white patch in wing. Best distinguished by size and habitat. **Voice** Song is similar to Siberian Stonechat but slower, deeper and richer. Call is a metallic *tek tek*. **Habitat** Nests in wet meadows in rocky alpine and subalpine slopes. **Behaviour** Chooses prominent perches from which it drops to ground to seize insect prey. Nests in rock crevices. **Status** Very rare and local breeding visitor in Mongol-Altai and Khangai mountain ranges, and possibly many other unexplored high-altitude sites; rare passage migrant elsewhere, late April to early September. **Conservation** Vulnerable globally and Near Threatened in Mongolia due to habitat loss and degradation. Listed in the Threatened Birds of Asia (2001) and Mongolian Red Book (2013). [Alt: Hodgson's Bush Chat]

Common Rock Thrush *Monticola saxatilis* 18–19 cm

ID The notably short-tailed, large-headed, ground-haunting rock thrushes are unlikely to be mistaken for other thrush-like species in Mongolia. Breeding male of this species is closest to breeding male White-throated Rock Thrush, which see for distinctions. Like other rock thrushes, adult female, imm and juv are brownish and scaly on head and underparts, but have trace of rufous on sides and show rufous (vs. dark) tail in flight. **Voice** Song (sometimes in display flight) is a rather short series of melodious phrases, often with abrupt ending. Call is a low *chak* or *chak-chak*, and whistled *piu*. **Habitat** Dry rocky mountain slopes with sparse vegetation. **Behaviour** Typically rather wary and unapproachable, though often chooses prominent perches. Flicks tail often. Nests in cavities under rocks, in walls or stone ruins. **Status** Uncommon breeding visitor in all major mountain ranges countrywide and uncommon passage migrant throughout, late April to late August (early September in the south). [Alt: Rufous-tailed Rock Thrush]

Blue Rock Thrush *Monticola solitarius* 17–19 cm

ID Overall dark blue coloration of adult male, together with typical rock thrush form and behaviour, are diagnostic. The race in Mongolia, *M. s. philippensis*, has mainly dark rufous underparts. Adult female and imm are darker overall than other rock thrushes and have plain (vs. scaly) crown and back. This species also has longer bill and tail than other rock thrushes. **Voice** Song is similar to Common Rock Thrush's, melodic but less soft, more chirpy. Call is a Eurasian Nuthatch-like *hwit hwit*; also, low *chak-chak* and thin *seeee*. **Habitat** Dry rocky mountain slopes with cliffs, typically nesting at higher elevation than Common Rock Thrush. **Behaviour** Similar to Common Rock Thrush. **Status** Very rare and local breeding visitor to Khavirga and Ih Khyangan, and Nömrög River valley, Dornod province; also recorded in Herlen River valley, Dornod province, on migration. Presumably present late April to late August. **Taxonomy** The local form is sometimes considered to be a separate species, Eastern Blue Rock Thrush.

White-throated Rock Thrush *Monticola gularis* 14–15 cm

ID Most likely to be confused with Common Rock Thrush, but adult male has white (vs. blue) throat, black facial mask and back, prominent white wing-patch, and dark (vs. rufous) tail. Adult female has olive (vs. warm brown) plumage with white throat-stripe, and lacks rufous on scaly sides. **Voice** Song is a rather slow, melodious whistle, often described as 'melancholy', interspersed with short trills. Call is a hard *chak* and softer *tsip*. **Habitat** Open montane forest with brushy undergrowth on rocky slopes. **Behaviour** Nests in cavities in tree stumps or among roots, as well as rock crevices; otherwise similar to Common Rock Thrush. **Status** Very rare and local breeding visitor to Khalkh and Nömrög Rivers, Dornod province and rare passage migrant in eastern Mongolia, Presumably late April to early September. A single bird documented in planted poplar trees in Mörön plantation, Hentii province, 5 September 2013.

1st-winter

Siberian Stonechat

maurus ♂

White-throated Bush Chat

♂

♀

stejnegeri ♂

♂

♀

♂ non-breeding

♂ breeding

Common Rock Thrush

♀ breeding

♀ breeding

♂

♀

♂

♂ breeding

philippensis

♂

♂ breeding

Blue Rock Thrush

♂ 1st-year

♂ breeding

White-throated Rock Thrush

Northern Wheatear *Oenanthe oenanthe* 13–14 cm

ID Breeding adult male unmistakable, with combination of grey upperparts, black mask and wings, buffy throat, and typical wheatear T-mark on tail. Adult female is similar but duller and lacks the mask. Non-breeding male is browner overall but retains basic spring pattern. Adult female and imm may be confused with Isabelline Wheatear, which see for distinctions. **Voice** Song is an explosive combination of cackles, warbles and whistles, sometimes including mimicry of other birds, typically delivered from top of rock or in short song-flight. Call is a hard *chlak*; also a sharp whistle, *wheet!* **Habitat** Nests in wide variety of montane habitats typically dominated by cliffs or rocky slopes, but occurs in other open habitats on migration. **Behaviour** Nests in abandoned burrows of small rodents, rock crevices and artificial structures. Chases arthropod prey in energetic runs. Not shy. **Status** Common or very common breeding visitor throughout Mongolian highlands, and fairly common passage migrant in all regions, late April to late August (early September in the Gobi).

Pied Wheatear *Oenanthe pleschanka* 14–15 cm

ID Striking black-and-white plumage pattern of adult male is unmistakable. Adult female and imm also readily distinguished from other Mongolian wheatears by their dirty grey-brown plumage and absence of eyebrows. In flight, all plumages show prominent narrow T-pattern with black also extending up sides of tail and white lower back as well as rump. Note: White-throated '*vittata*' form occurs in Gobi-Altai and Mongol-Altai mountain ranges. **Voice** Song consists of short, garbled phrases containing varied warbles and chirps and mimicry of other birds. Usual call is a double clicking *zak-zak*; also shrill *tsweep*. **Habitat** Cliffs and other rocky terrain, including forest edges in mountains with lush or sparse shrubbery, as well as other open habitats on migration. **Behaviour** Nests in rock crevices. Flycatches from top of rocks or other prominent perches. Hops rather than runs when foraging on ground. **Status** Uncommon, but widely distributed breeding visitor in montane habitats, occurring from extreme west to east and north to south. Uncommon passage migrant throughout, late April to late August (early September in the south).

Desert Wheatear *Oenanthe deserti* 13–14 cm

ID Adult male unmistakable by combination of black throat and sandy-brown upperparts. In all plumages, nearly all-black tail (base of tail is white) is diagnostic, especially in flight but also while perched. In addition to tail, adult female and imm can be distinguished from Isabelline, Pied and imm Northern Wheatears by more compact form and habitat preference. **Voice** Song is a short, plaintive (even mournful) fluting, punctuated with short rattle or trill. Calls include *chuk*, *krrr* and *te-oo*. **Habitat** Rocky areas and dry river beds with scattered low shrubs in arid steppe and hardpan desert; usually avoids pure sands and high elevation. Often attracted to herders' camps in Gobi Desert. **Status** Fairly common breeding visitor from Great Lakes Depression and Mongol-Altai range south through Gobi-Altai, becoming increasingly common further south, and uncommon passage migrant throughout its range, late April to late August (early September in the Gobi).

Isabelline Wheatear *Oenanthe isabellina* 16–18 cm

ID Easily confused with imm Northern Wheatear but slightly larger, with longer legs and conspicuous habit of standing up almost vertically. Up close, eyebrow in front of eye is white, not buff. Sexes alike, but male shows blacker lores. In flight, black terminal tail-band is broader than in Northern and T-'stem' thus shorter. Tolerates open, drier and short-vegetated habitats better than Northern. **Voice** Song, often in flight, tends to be more prolonged and 'conversational' than Northern's but with similar combination of cackles and whistles (including diagnostic 'wolf-whistles') and imitations of other birds and mammals. Call is a domestic chick-like *cheep*; also *chip* and *chack*. **Habitat** Typically nests in dry, open habitats with short vegetation including scattered low bushes or on rocky slopes of mountains, but also in steppe at margins of taiga forest. Occurs in all open habitats on migration. **Behaviour** Builds nest in burrows of voles, ground squirrels, marmots, pikas and hares. Usually seen foraging on ground, though will also perch on wires and other lookout points. **Status** Very common breeding visitor, and fairly common passage migrant throughout, late April to late August.

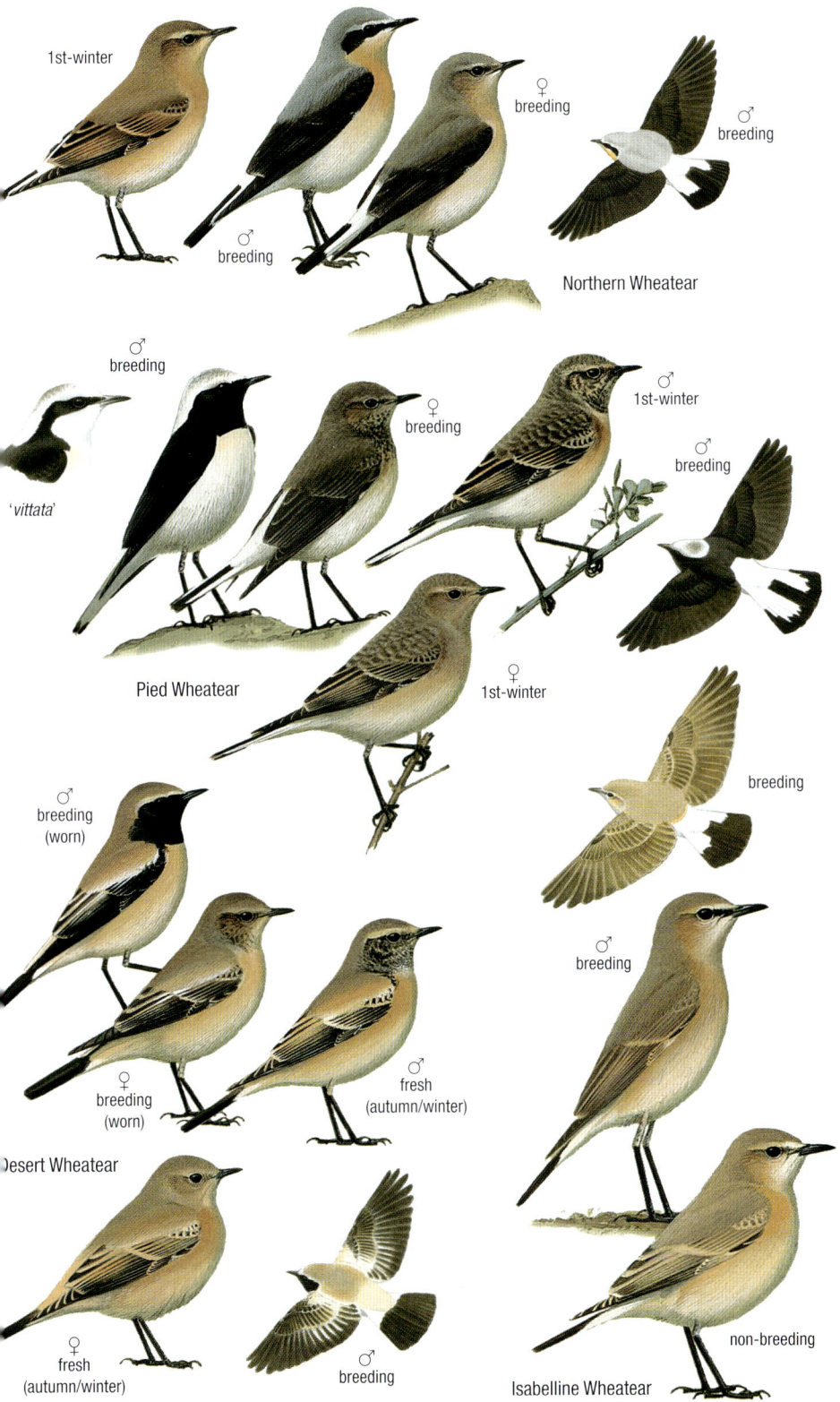

1st-winter

♂
breeding

♀
breeding

♂
breeding

Northern Wheatear

♂
breeding

'vittata'

♀
breeding

♂
1st-winter

♂
breeding

Pied Wheatear

♀
1st-winter

breeding

♂
breeding

♂
breeding
(worn)

♀
breeding
(worn)

♂
fresh
(autumn/winter)

Desert Wheatear

♀
fresh
(autumn/winter)

♂
breeding

non-breeding

Isabelline Wheatear

European Pied Flycatcher *Ficedula hypoleuca*   13 cm

ID Bold black-and-white plumage of adult male is unmistakable in Mongolia. Plainer adult female and imm are grey on head and mantle, but black-and-white wing pattern remains diagnostic. **Voice** Song is a simple phrase of short, sweet warbles and chirps. Calls include a brief *vit* and *weet*. **Habitat** Mainly deciduous woodlands, often near water and in gardens on migration. **Behaviour** Like Spotted and other flycatchers; flicks wings and tail when calling. **Status** Vagrant. Sight records as follows: two near Tolbo Nuur, Bayan-Ölgii province, 1 June 2006; single birds at Khovd town, Khovd province, 22 September 2006 and 10 May 2007; and four at latter site, 9 and 14 October 2007.

Yellow-rumped Flycatcher *Ficedula zanthopygia*   12–13 cm

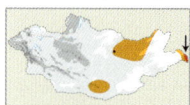

ID Striking black, yellow and white plumage of adult male can only be mistaken for that of Narcissus Flycatcher, which differs in its yellow (vs. white) supercilium, its orange (vs. golden-yellow) throat and its less extensive white wing-patch. Generally plain adult female retains yellow rump and vestige of white wing-patch, both features lacking in Narcissus. **Voice** Song is a three- or four-note thrush-like warble with burred accents. Call is a rather hard and rattling *prrrt*. **Habitat** Open deciduous woodland and willow copses in river valleys. **Behaviour** Flycatches from tree-tops or lower, open perches. **Status** Very rare and local breeding visitor in Khalkh and Nömrög River valleys, Dornod province, and rare passage migrant in Terelj, upper Tuul and Onon River valleys, Hentii and Töv provinces, late April to late August (early September in the east). [Alt: Tricolor Flycatcher]

Narcissus Flycatcher *Ficedula narcissina*   11–13 cm

ID Likely to be mistaken only for Yellow-rumped Flycatcher. Adult male distinguished by yellow (vs. white) supercilium, bright orange (vs. yellow) throat and upper breast and less extensive white in wing. Adult female lacks contrasting yellow rump. **Voice** Song varies but typically a two- or three-note clear whistle, ending with high lisping note; may include mimicry. Calls include a bubbling *brrrt*, quiet *tink-tink* and plaintive, repeated *piu-piu-piu* notes. **Habitat** Deciduous woodland of willow, birch and poplar in river valleys. **Behaviour** Similar to Yellow-rumped Flycatcher. **Status** Vagrant. Known from one specimen of breeding male in Nömrög River valley, Dornod province, 25 May 1998.

Mugimaki Flycatcher *Ficedula mugimaki*   12–13 cm

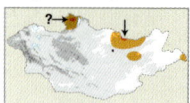

ID Similar in form and general plumage pattern to Yellow-rumped and Narcissus Flycatchers, but adult male readily distinguished by rufous-orange (vs. yellow or orange) throat and breast and short white stripe behind (vs. over) the eye. Might at first be taken for redstart or Brambling, but compare behaviour and plumage details. Adult female is drab but still distinguishable by paler rufous-orange (vs. whitish) underparts and warmer brown coloration above. **Voice** Song is a loud and emphatic series of sharp, accelerating whistled notes ending in an abrupt twitter. Bubbling *berrriit* call is similar to that of Narcissus Flycatcher; also, other soft and rattling calls. **Habitat** In Mongolia, known to nest in mixed montane taiga forest with old trees. Visits willow copses in river valleys during migration. **Behaviour** Similar to Yellow-rumped Flycatcher. **Status** Very rare and local breeding visitor to Onon River valley and possibly Darkhad Depression; also, sight records in Khalkh and Nömrög River valleys during breeding season, late April to late August.

Taiga Flycatcher *Ficedula albicilla*   11–12 cm

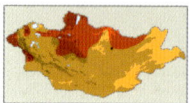

ID Adult male readily distinguished from other Mongolian flycatchers by combination of lead-grey head, rufous-orange throat and black-and-white (wheatear-like) tail pattern. Plain, unstreaked female and imm birds told from similar Asian Brown Flycatcher by tail pattern. See also behaviour. **Voice** Song begins with several high-pitched sweet notes and accelerates into richer descending chirps. Call is a hard trilling *trrrrt*, single or double *chick*, etc. **Habitat** Nests in coniferous, deciduous and mixed montane forest. Occurs in almost every other Mongolian habitat on migration except alpine zone. **Behaviour** Like other flycatchers, often catches insects in aerial sorties from a fixed perch, but also feeds actively within canopy like a warbler; frequently cocks its tail. **Status** Fairly common breeding visitor throughout mountain ranges of northern Mongolia and common passage migrant throughout country, often in numbers, late April to late September. **Taxonomy** Formerly considered a subspecies of Red-breasted Flycatcher *F. parva*. [Alt: Red-throated Flycatcher]

1st-winter

♀

♂

Yellow-rumped Flycatcher

♀
breeding

♂
breeding

European Pied Flycatcher

♀

♂

Narcissus Flycatcher

♂
1st-winter

♀

♂

♀

♂

Mugimaki Flycatcher

Taiga Flycatcher

Spotted Flycatcher *Muscicapa striata* 13–14 cm

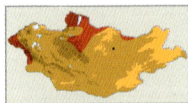

ID Distinguished from similar Dark-sided and Asian Brown Flycatchers (with which it often co-occurs) by its larger size, slimmer form and proportionally large head; also, by fine but distinct streaking on forecrown, throat and breast. Spotted also lacks the smaller species' dark malar stripe and prominent pale eye-ring. **Voice** Song is a soft, squeaky combination of *tzee* and *chup* notes, with long pauses between phrases. Calls include *zeee* or *zeee-zuk* in alarm; also, a hard *chick*. **Habitat** Nests in open conifer (larch) forest and rarely mixed woodlands. Frequents many other habitats on migration. **Behaviour** Sallies out and back to same perch in flycatching sorties. Frequently flicks wings and tail. **Status** Uncommon breeding visitor at moderate elevations in Mongol-Altai, Hövsgöl, Khangai and Hentii ranges and uncommon passage migrant throughout, late April to mid-September.

Dark-sided Flycatcher *Muscicapa sibirica* 12–13 cm

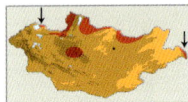

ID Best distinguished from similar Spotted and Asian Brown Flycatchers by heavier, smudgy breast streaks; sooty-grey flanks; and absence of head streaking (cf. Spotted). Sexes alike. **Voice** Song is a soft jumble of thin, high-pitched phrases, followed by more melodious trills and whistles. Calls include a thin, high-pitched *tswee*, *swit* or *cheeer*. **Habitat** Nests in deciduous, coniferous and mixed montane forest, typically with tall old trees. On migration, occurs in other wooded habitats, including parks and gardens. **Behaviour** Similar to Spotted and Asian Brown Flycatchers. **Status** Uncommon breeding visitor in Great Lakes Depression and Hövsgöl, Khangai and Hentii mountain ranges and possibly in Buir Lake-Khalkh River-Khyangan Region (Dornod province) and uncommon to fairly common passage migrant throughout Mongolia, late April to early September. [Alt: Sooty Flycatcher]

Grey-streaked Flycatcher *Muscicapa griseisticta* 12.5–14 cm

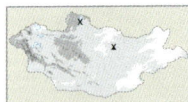

ID Most closely resembles Dark-sided Flycatcher, but slightly larger with longer bill and tail. Also, breast streaks bolder and more clearly defined, and has distinct whitish feather edgings in wings (primary-coverts and tertials). **Voice** Song is a sharp, high-pitched twitter. Call is a soft *hseeest* and other soft, thin notes. **Habitat** Woodlands, including wooded parks. **Behaviour** Similar to Spotted and other flycatchers. **Status** Vagrant. One at Davaany Zörlög, 10 km west of Ulaanbaatar, Töv province, 11 June 1990 and one in Hövsgöl Mountain in 1992.

Asian Brown Flycatcher *Muscicapa dauurica* 11–13 cm

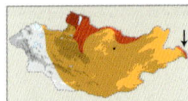

ID Best told from similar Spotted and Dark-sided Flycatchers by its plain, 'clean' appearance without streaking on head and breast (a few individuals have very fine breast streaks). Spotted also lacks its prominent eye-ring and dark malar stripe. Seen from below, its notably broad bill with yellow-pink base is further distinction. **Voice** Song is similar to that of Dark-sided Flycatcher but slower-paced. Calls include a grating *trrrr*; also soft *chuck* and *swee* notes. **Habitat** Open deciduous (rarely coniferous) montane forests; many other wooded habitats on migration. **Behaviour** Similar to Spotted Flycatcher. **Status** Fairly common breeding visitor in river valleys in Hövsgöl, Khangai and Hentii ranges and highlands of Buir Lake-Khalkh River-Khyangan Region, Dornod province, and common passage migrant throughout Mongolia, early May to late August (early September in the steppe and Gobi).

Spotted Flycatcher

adult worn

adult fresh

1st-winter

adult

Dark-sided Flycatcher

adult worn

adult fresh

Grey-streaked Flycatcher

adult worn

adult fresh

Asian Brown Flycatcher

House Sparrow *Passer domesticus*   14–16 cm

ID Adult male easily recognised by its combination of rich chestnut back and nape, grey crown, white cheeks and black throat and upper breast. Adult female and juv much duller, but combination of tan eye-stripe, single bold white wing-bar, and dull greyish-brown breast (plus usual proximity of males) are distinctive. See also female Saxaul Sparrow. **Voice** Rather noisy. Call is a strong *chirrup* and shorter *chip*. Song is a simple series of loud chirps. **Habitat** Typically nests in and around human habitations, including in major cities. **Behaviour** Gregarious, feeding in flocks and often nesting in loose colonies in cavities of artificial structures. Feeds on seeds of various plants and any edible scraps in urban areas, as well as insects, especially when breeding. **Status** Very common resident breeder throughout Mongolia, though often absent from wilder regions with fewer human inhabitants.

Saxaul Sparrow *Passer ammodendri*   14–15 cm

ID Sandy grey-brown upperparts, distinctive head pattern with black crown and nape, and rufous patch above and behind eye distinguish adult male from other *Passer* sparrows in Mongolia. Female and juv similar to House Sparrow but with paler upperparts and finer black streaks on back. **Voice** Call consists of simple chirps, some higher and more ringing than others. Song is a series of call-like notes, softer and more melodious than House Sparrow's. **Habitat** Mature stands of Saxaul trees (*Haloxylon ammodendron*) in Gobi Desert. **Behaviour** Nests in holes in old Saxaul trees or in stick nests of raptors and ravens, sandy cliff crevices or artificial structures. Post breeding, forms flocks of up to 60 individuals and forages for seeds, especially of poplar and tamarisk. Rather shy when nesting but in winter visits camps of herders to scavenge seeds from animal droppings and waste grain. **Status** Uncommon to locally fairly common resident breeder, essentially confined to Gobi Desert. Two races occur: *P. a. nigricans* in SW Mongolia and *P. a. stoliczkae* in S Mongolia. **Conservation** Least Concern globally, but Near Threatened in Mongolia due to restricted range of subspecies *P. a. stoliczkae* and destruction and degradation of Saxaul trees. Listed as Near Threatened in the Mongolian Red Book (2013).

Eurasian Tree Sparrow *Passer montanus*   12.5–15 cm

ID Adult is readily distinguished from similar House Sparrow by its chestnut (vs. grey) crown and prominent black cheek-spot. Sexes are alike. Juv is like adult but duller and less distinctly marked. **Voice** Song like House Sparrow's but a little higher-pitched. Call is a House Sparrow-like chirping; flight call is somewhat metallic *chep*. **Habitat** Breeds in deciduous forest along river valleys and other wild habitats with trees much more readily than House Sparrow, but also thrives in urban areas where it often outnumbers the latter. **Behaviour** Much like House Sparrow with which it often associates, especially in cities. Nests in cavities in trees and buildings. **Status** Very common to locally abundant resident breeder throughout Mongolia; even more widespread than House Sparrow.

Rock Sparrow *Petronia petronia*   14–17 cm

ID Adult resembles large, chunky female House Sparrow but distinguished by heavy breast streaking, much heavier bill with yellowish lower mandible, pale central crown-stripe and yellow throat patch (often obscure). Also, see habitat. Juv very like adult but with darker brown stripes on head, yellow gape, pinkish lower mandible, and lacks yellow throat-patch. **Voice** Chirps are more musical than House Sparrow's; also gives distinctive wheezing *pee-yip*. Song comprises repeated calls. **Habitat** Rocky mountains and hillsides in all types of steppe and desert; also human settlements. **Behaviour** Gregarious, especially after breeding, often flocking with other sparrows and finches. Nests in rock cavities, stick nests of raptors and corvids, and artificial structures. **Status** Common to very common resident breeder throughout Mongolia. Subspecies in Mongolia is *P. p. brevirostris*, but *P. p. intermedia* may occur in SW.

House Sparrow

stolickzae

♂

nigricans

♀

Saxaul Sparrow

juvenile

adult

Eurasian Tree Sparrow

intermedia

adult

brevirostris

adult

Rock Sparrow

White-winged Snowfinch *Montifringilla nivalis* 16.5–19 cm

ID Bold, patchy black/white/greyish-brown colour pattern highly distinctive in all plumages, especially in flight. Non-breeding birds and juv have yellow bill and indistinct or no black bib. Similar Snow Bunting has prominent rufous patches on head, breast and back while in Mongolia in winter, and distribution of the two species overlaps only in extreme north-west. **Voice** Song, often given in parachuting display-flight, is a repeated *see-tut-chur see-tut-chur*. Call is a throaty *zweek* or rolling *chrrrrt*. **Habitat** Rocks and cliff faces near springs and creeks in high mountains (2,200–4,000 m); descends to lower slopes in winter. **Behaviour** Active and gregarious. Forages on ground in flocks and nests in rock cavities. Scavenges for grain near nomad camps in winter. **Status** Locally common resident breeder in drier mountain ranges from Mongol-Altai through Gobi-Altai. However, distribution is extremely erratic and species is absent in many areas which appear to be suitable.

Père David's Snowfinch *Pyrgilauda davidiana* 12–14 cm

ID Adult male's combination of small size, sandy-brown and white coloration and black throat and forehead are diagnostic in Mongolia. Adult female is very similar but with reduced head markings. Juv lacks black head markings, but shows prominent white wing-stripe in flight. **Voice** Song and calls are simple and repetitive chirps. **Habitat** Open steppe with short-vegetation near water, descending to lower valleys in mountainous areas outside of breeding season. **Behaviour** Often nests in rodent and pika burrows. Forms large feeding flocks in winter and scavenges around nomad camps; often quite tame. **Status** Fairly common resident breeder across Mongolia; absent only in forested areas and Gobi. **Taxonomy** Formerly placed in genus *Montifringilla*. [Alt: Small Snowfinch]

Alpine Accentor *Prunella collaris* 15–17 cm

ID Largest and stoutest Mongolian accentor, most closely resembling Altai Accentor with which it co-occurs, but distinguished from that species by grey (vs. rufous streaked) breast, fully and densely barred throat (vs. white with black border), and yellow-based (vs. mainly black) bill. Juv. is similar though more obscurely marked. **Voice** Song is a brief melodious warbling. Call is a brief *kya* call; usual flight call is a loud lark-like *chirrp*. **Habitat** Nests in rocky alpine meadows in high mountain ranges. Occurs at lower elevations in rocky mountains of steppe and desert steppe zones during seasonal movements. **Behaviour** Forages on the ground for invertebrate prey and seeds in winter. Often rather tame. **Status** Uncommon resident breeder in alpine zone of Mongol-Altai, Gobi-Altai, Khangai, and Hentii ranges; altitudinal migrant.

Altai Accentor *Prunella himalayana* 16 cm

ID Similar to Alpine Accentor, but breast rufous-streaked, not uniform grey; throat white with black border, and not dense with black scales; and bill mainly black, not yellow at base. Juv is more obscurely marked, though still distinctive, but note pink bill base. **Voice** Sweet trills and warbles similar to other accentors. Call is a finch-like *tee-tee*. **Habitat** Nests in dense shrubbery in rocky alpine meadows; in winter, descends to lower rocky and grassy hillsides. **Behaviour** In winter, forms foraging flocks of 6–30 individuals. **Status** Fairly rare and local resident breeder in alpine and subalpine zones of Mongol-Altai, Hövsgöl, Khangai, Hentii and Gobi-Altai ranges. [Alt: Rufous-streaked Accentor, Himalayan Accentor]

adult

White-winged Snowfinch

♂

Père David's Snowfinch

juvenile

Alpine Accentor

adult

juvenile

Altai Accentor

adult

Brown Accentor *Prunella fulvescens*   15 cm

ID Might at first be mistaken for warbler or chat, but distinguished by posture and foraging behaviour, which are more bunting-like, and by face pattern. Similar to Siberian Accentor, but distinguished by white supercilium, olive-brown (vs. warm brown) upperparts, and buffish (vs. light rufous) belly. Black-throated Accentor differs by its black-throat. Juv is more obscurely marked, with streaky breast and indistinct face mask. **Voice** Song is short, high-pitched and warbling. Usual call is *zyet zyet zyet*. **Habitat** Nests in thickets and juniper scrub on dry rocky slopes in alpine and subalpine zones. Nest may be in rock crevices or small trees. In winter, moves to lower elevation rocky hills with thick brush in steppe and desert zones. **Behaviour** Feeds on ground and in low shrubbery. Fairly quiet and unassuming, but also rather tame. Forms small foraging flocks in winter. **Status** Uncommon to fairly common resident breeder in all major mountain ranges, and occurs countrywide post-breeding, except for driest areas of south.

Siberian Accentor *Prunella montanella*   13–14 cm

ID Similar to Brown Accentor but more boldly and brightly marked, with yellow-ochre supercilium, throat and upper breast; reddish-brown streaks on back and sides; grey nape collar; and pale ear-spot on black facial mask. **Voice** Song is a short, shrill warble. Call is a high-pitched three- or four-syllable call, *ti-ti-ti* or *si-si-si*. **Habitat** Breeds in subarctic birch and willow thickets, bog edges and open conifer woods. In Mongolia, migrants visit a wide range of habitats from forest to desert. **Behaviour** Similar to other accentors, though rather shy, often retreating to cover. Flicks wings while feeding. **Status** Uncommon passage migrant throughout Mongolia, mid-April to early May and late August to mid-September. Nesting in far north is possible but so far unconfirmed.

Black-throated Accentor *Prunella atrogularis*   13–14 cm

ID Breeding adult is similar to Brown and Siberian Accentors, but black throat is diagnostic. In juv, throat is scaly but still distinctive. **Voice** Song is a fairly short, rapid pastiche of twitters and chirps. Call is a high *ti-ti-ti*, similar to other accentors. **Habitat** Nests in low, dense scrub among rocks near coniferous forest in subalpine zone. Occurs in brushy areas at lower elevations during migration. **Behaviour** Similar to Brown Accentor. **Status** Very rare, very local and little-known breeding visitor to Khovd and Yolt River valleys in Mongol-Altai range; also recorded as rare passage migrant in Great Lakes Depression, late April to late August (early September in the Gobi).

Kozlov's Accentor *Prunella koslowi*   15 cm

ID The most distinctive plumage characteristic of this species is that it has none! It is easily recognised by its overall light tan coloration, which – on close inspection – shows a few obscure streaks on the back and spots on the breast. All other Mongolian accentors are spectacular by comparison. **Voice** Song is a fairly terse burst of alternating chirps and twitters much like that of other accentors. Call is a series of high-pitched, dry notes. **Habitat** Nests in thickets of juniper, *Caragana* and other shrubs up to 1 m high in Mongolia's drier mountains. After breeding it descends to brushy river valleys on lower slopes. **Behaviour** Forms post-breeding flocks of 6–30 individuals. Known to enter towns in winter. **Status** Fairly rare and local resident breeder, mainly in drier mountains of west and south, from Great Lakes Depression through the Gobi-Altai range. Non-breeding birds also recorded from Khalkh River, Dornod province. **Conservation** This species is near-endemic in Mongolia, which comprises the vast majority of its range. Despite its limited distribution, its population is believed to be large and stable. [Alt: Mongolian Accentor]

juvenile

adult

Brown Accentor

dull individual

breeding

Siberian Accentor

1st-winter

adult
worn

Black-throated Accentor

adult

Kozlov's Accentor

Yellow Wagtail *Motacilla flava*

15–16 cm

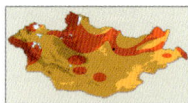

ID With a slighter build and shorter tail than other Mongolian wagtails, this species also has distinctly olive to brownish upperparts in all plumages, unlike its congeners. Plainly marked greyish juv and first-winter birds might be confused with young Citrine Wagtail, but in addition to build and coloration have less conspicuous wing-bars, a pale (vs. black) lower mandible and pale yellow undertail-coverts. Five different subspecies of Yellow Wagtail occur in different parts of Mongolia in different seasons. Adult and especially adult male of these races can be readily identified by their head pattern and coloration, though all hybridise where their ranges overlap.

M. f. tschutschensis Adult male has grey head with narrow white supercilium and dark ear-coverts. Throat yellow, some with white on malar.

M. f. leucocephala Adult male has white head with variable grey wash on crown and ear-coverts. Throat yellow with white on malar.

M. f. macronyx Adult male has grey head and yellow throat, usually without any supercilium.

M. f. beema Adult male has blue-grey head with broad white supercilium and white area on ear-coverts below the eye. Throat yellow with whitish chin.

M. f. taivana Adult male is readily distinguished from other forms in Mongolia by combination of olive crown, yellow supercilium and black mask. Adult female is similar but mask is not as dark.

Voice Song, delivered from elevated perch, is a very simple, dry, two- or three-note *tsre-tsree* or *tse-tse-o*. Call is a piercing *tsweeep*. **Habitat** Wet meadows, bog and marsh edges, and waterlogged river banks and lake shores. **Behaviour** Nests in hollows or among thick, low vegetation on ground; often associates with livestock. Forms post-breeding flocks of up to 200 birds. **Status** Fairly common breeding visitor, mainly across northern half of country, and uncommon passage migrant throughout, late April to early September. Of five subspecies, *M. f. tschutschensis* occurs in Hövsgöl region; *M. f. leucocephala* in Great Lakes Depression; *M. f. macronyx* in the east and south of the country; and the other two only as passage migrants (*M. f. beema* from north and west; and *M. f. taivana* from far east – the latter has been recorded in scattered localities in Mongolia, including Böhög River in eastern Hentii range; and several localities, including one record of five birds, from wetlands in Dornod province in August 2004 and May 2009). **Taxonomy** The various forms are sometimes considered to be two separate species: Western Yellow Wagtail *M. flava* (comprising *leucocephala*, *beema* and other extralimital races) and Eastern Yellow Wagtail *M. tschutschensis* (comprising *tschutschensis*, *macronyx* and *taivana*).

Citrine Wagtail *Motacilla citreola*

15–17 cm

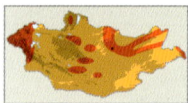

ID Adult male unmistakable with glowing yellow head and breast, slate-grey back, white wing-bars, and black neck 'shawl'. Adult female is much duller and might be mistaken for adult female Yellow Wagtail but retains grey (vs. olive) back and has very bold white wing-bars. Imm (1st-winter) lacks all trace of yellow but retains diagnostic grey back and white wing-bars, and has distinctive pale margin surrounding a dark cheek patch. **Voice** Song is a brief, sharp *tzreeep tchip-tchip*. Call is an explosive, harsh *tzreep*. **Habitat** Bogs, wet meadows, tussock marshes and soggy shores of rivers and lakes with marginal shrubs. **Behaviour** Nests on ground in hollows in earth near water, sheltered by overhanging rocks or shrubbery. **Status** Fairly common breeding visitor in appropriate wetland habitat, mainly in highlands across Mongolia; absent from driest parts of extreme south. Fairly common passage migrant throughout. There are two records of the Black-backed race *M. c. calcarata*: one near Choibalsan in the east, and one from Khovd in the west. Flocks with Yellow Wagtails on migration, late April to late August (early September in the Gobi). [Alt: Yellow-headed Wagtail]

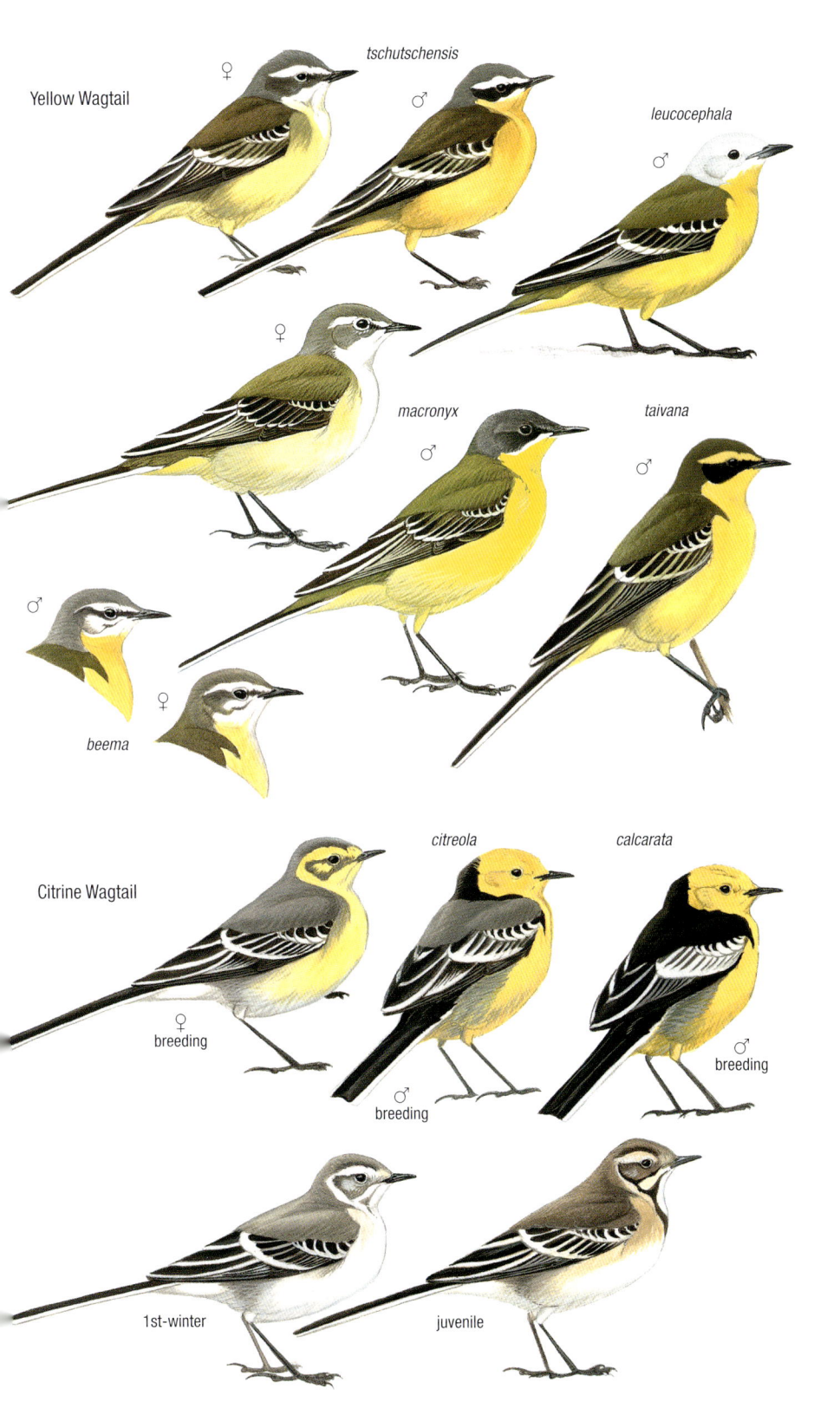

Yellow Wagtail

♀ *tschutschensis* ♂

leucocephala ♂

♀ *macronyx* ♂ *taivana* ♂

♂
beema ♀

Citrine Wagtail

citreola *calcarata*

♀
breeding

♂
breeding

♂
breeding

1st-winter juvenile

Grey Wagtail *Motacilla cinerea* 17–20 cm

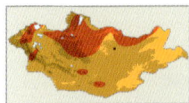

ID Adult male is only wagtail to combine blue-grey head and back, bright yellow underparts and black throat. Adult female and juv are paler yellow to whitish below and lack black throat, but retain distinctive grey upperparts and bright yellow undertail-coverts. This is the most elegant wagtail, with exceptionally long tail. See also habitat and behaviour. **Voice** Song is a complex and varied series of trills and whistles. Call is a loud, sharp and explosive *tsit-tsit*. **Habitat** Large, fast-flowing rivers and lake shores with heavy wave action. **Behaviour** Typically forages actively at water's edge, often perching on stones in midstream and darting out to catch flies. Bobs its tail more constantly and energetically than other wagtails. Nests in rock crevices near water. **Status** Fairly common breeding visitor in highland rivers and lake shores, mainly in western and north-central Mongolia, but also in Gobi-Altai and fairly common passage migrant throughout, late April to late August (early September in the Gobi).

White Wagtail *Motacilla alba* 16–19 cm

ID In most plumages easily distinguished from other Mongolian wagtails by bold black, white and grey plumage pattern, with no trace of yellow. Juv resembles grey juv Citrine and Yellow Wagtails, but typically shows trace of adult's black breast-patch. See also distinctive face pattern of juv Citrine. Four subspecies occur in Mongolia, which – especially in adult plumages – can be distinguished by variations in plumage (see plate). These races are arranged roughly by geographical distribution and status, with *M. a. personata* (sometimes considered a full species, Masked Wagtail) occurring mainly in western highlands; *M. a. baicalensis* across the country; *M. a. leucopsis* in south-east; and *M. a. ocularis* as passage migrant. All these forms intergrade where their ranges overlap. **Voice** Song is a complex, excited twittering, incorporating chirps and warbles, or much simpler *tsu-su tsu-su zwee zwee*. Call is a sharp high *chissik* or *tslee-vit*. **Habitat** Nests near all types of wetlands in rock crevices, tree-holes and artificial structures. However, migrants occur in all aquatic habitats, including desert oases. **Behaviour** Walks quickly with dancing gait, constantly bobbing tail. Flycatches and picks up food from ground. **Status** Common breeding visitor and passage migrant throughout Mongolia, late April to mid-September. Post-breeding flocks may number up to 150 individuals.

Forest Wagtail *Dendronanthus indicus* 16–18 cm

ID Unmistakable in all plumages due to its unique combination of olive-brown upperparts; double black breast-bands; and pink bill, legs and feet, along with typical wagtail form and behaviour. **Voice** Song is a series of high-pitched squeaky notes; has been compared to Great Tit song. Call is a single or double ringing metallic *pink*; also a softer *tsip*. **Habitat** Typically, open glades and paths in forests. **Behaviour** Feeds on ground, well-camouflaged in forest leaf-litter, flying into trees when disturbed. Walking birds commonly sway rear and tail from side to side. **Status** Vagrant from east or south. Two records of single birds at Juulchin Gobi tourist camp, Ömnögobi province, June 2000 and 2002; and Terelj, Töv province, August 2006.

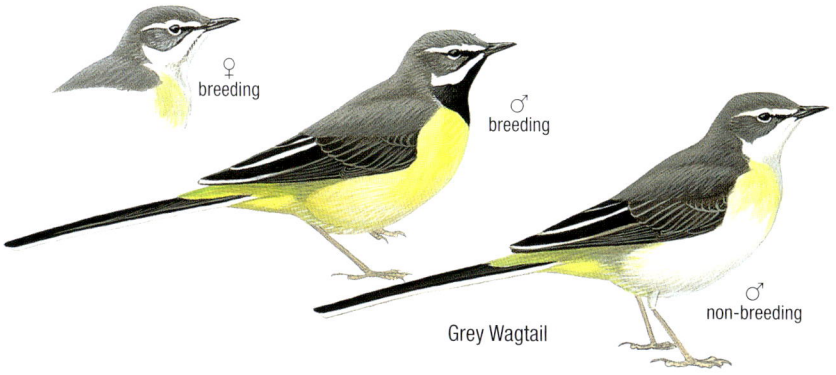

breeding ♀

♂ breeding

♂ non-breeding

Grey Wagtail

juvenile

personata

♂ breeding

leucopsis

♂ breeding

♀

baicalensis

♂ breeding

♀ 1st-winter

White Wagtail

ocularis

♂ breeding

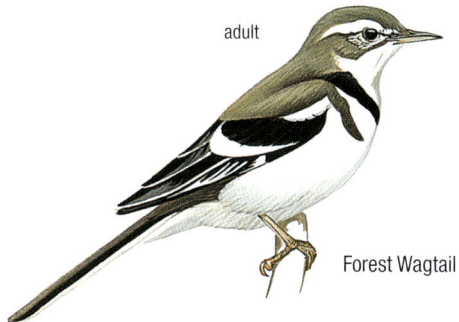

adult

Forest Wagtail

PLATE 99: PIPITS I

Richard's Pipit *Anthus richardi*   17–20 cm

ID Notably larger than other Mongolian pipits with distinctively long tail and legs; long, heavy bill; and characteristic upright stance. Prominently streaked on head, back and breast. Exceptionally long hind-claws are sometime visible on perched birds at close range. Most likely to be confused with Blyth's Pipit, which see for further distinctions. See also Tawny Pipit and Eurasian Skylark. **Voice** Song is a simple, toneless, buzzy *tsii, tsii,tsii, tsii...*, usually given in circling, undulating song-flight. Call is a loud, harsh *schreep*. **Habitat** Nests in relatively rich, often moist grasslands. On migration, occurs in open habitats throughout, including arid steppe and desert. **Behaviour** When foraging on ground, runs and pauses, assuming upright, wheatear-like stance. Wary, flies some distance when flushed. **Status** Common breeding visitor in the mountain ranges across northern Mongolia and fairly common passage migrant throughout, late April to early September.

Blyth's Pipit *Anthus godlewskii*   15–17 cm

ID Very like Richard's Pipit, and sometimes not safely separable by plumage characters alone. But Blyth's is smaller and 'daintier' with shorter and thinner bill and comparatively shorter legs, tail, neck and hind-claws. Stance not as obviously upright as Richard's. Compare voice. **Voice** Song is very unlike that of Richard's and other pipits – a harsh *zret zret zret* phrase, ending with higher-pitched rattle. Call like Richard's but softer, less rasping, a rather nasal *psheeu*; also short *chet* notes. **Habitat** When nesting, tends to prefer drier, less moist steppe than Richard's, including rocky, sparsely vegetated mountain slopes, though not so dry as Tawny Pipit. Occurs in all open habitats on migration, including arid steppe and desert. **Behaviour** Similar to Richard's Pipit. **Status** Common breeding visitor as high as 3,000 m in all major mountain ranges across Mongolia and common passage migrant throughout, late April to early September. More widespread as a breeding species than Richard's Pipit.

Tawny Pipit *Anthus campestris*   16 cm

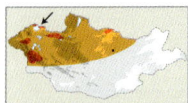

ID A large pipit, most likely to be confused with Richard's and Blyth's Pipits, but adult notably pale, sandy coloured and only faintly (vs. prominently) streaked on head, breast and back. Juv Tawny is boldly streaked, thus more like the other two similar species, but has shorter, thinner bill than Richard's and at close range shows dark stripe between eye and bill (lores), which Richard's and Blyth's lack. Compare also voice and habitat. **Voice** Song is a repeated two- or three-note *tseerluee. tseerluee...* usually given in undulating flight. Call is a slurred *chirleep* or *chelp*. **Habitat** Prefers drier habitats than Richard's or Blyth's, e.g. sandy riverbanks and lake shores; gravelly hillsides with sparse vegetation; and river valleys with scattered tall *Caragana* bushes and deciduous trees. Occurs in more diverse open habitats during migration. **Behaviour** Moves in wagtail-like manner, running with tail bobbing. Breeding adult male sings on top of bushes and small trees. **Status** Uncommon to rare breeding visitor in Great Lakes Depression and Mongol-Altai range to western Khangai and Hövsgöl ranges and Orkhon-Selenge River valley, and rare passage migrant throughout Mongolia, late April to late August (early September in the steppe and Gobi).

Buff-bellied Pipit *Anthus rubescens*   15–16 cm

ID Similar in form and general coloration to Water Pipit, with which it was once considered conspecific, but summer adult of subspecies recorded in Mongolia (*A. r. japonicus*) has prominently streaked (vs. almost clear) breast. The breast streaking of non-breeding birds and juv is blacker and more distinct than that of Water Pipit, creating possible confusion with Tree or Olive-backed Pipits. However, *japonicus* is stouter of build; distinctly buffy below with an unstreaked crown; has black legs; and has very obscurely streaked (vs. prominently streaked) back. Face pattern is also different, with prominent, complete eye-ring and no ear-spot. **Voice** A crisp, sharp *pit-it* and more drawn-out *tseep*. **Habitat** Nests in arctic and subarctic tundra. On migration, uses variety of open, often wet, habitats. **Behaviour** As Water Pipit. Flocks with Red-throated Pipit on migration. **Status** Rare but apparently regular passage migrant to far eastern Mongolia, late April to early May and late August to early September. [Alt: American Pipit]

Water Pipit *Anthus spinoletta*   15–17 cm

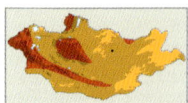

ID A rather stout, long-billed, alpine pipit. Breeding adult readily distinguished from most other Mongolian pipits by combination of pinkish underparts with minimal streaking on flanks; grey, unstreaked crown and nape; and brown back with dull streaking. However, see similar (much scarcer) Buff-bellied Pipit for distinctions. Autumn and winter birds develop heavy, smudgy streaking on breast and flanks. **Voice** Song (given in flight or from rock or low shrub) is a varied series of sharp, rapidly repeated notes that have been likened to a sewing machine, e.g. *tree-tree-tree, zru-zru-zru...* typically ending with distinctive, buzzy *zeeer, zeeer, zeeer*. Flight call is a high-pitched *weeeest*. **Habitat** Nests in alpine meadows, tundra and rocky slopes near water courses at high elevation. Visits other open, preferably wetland, habitats on migration. **Behaviour** Nests in and forages over open ground. **Status** Uncommon to fairly common breeding visitor to mountain tops in Great Lakes Depression and Mongol-Altai, Gobi-Altai, Khangai, Hövsgöl and Hentii ranges. Uncommon passage migrant throughout, late April to late August (early to mid-September in the Gobi).

juvenile

adult

Richard's Pipit

juvenile

adult

Blyth's Pipit

juvenile

1st-winter

adult

Tawny Pipit

breeding

non-breeding

Buff-bellied Pipit

japonicus

breeding

non-breeding

Water Pipit

blakistoni

Tree Pipit *Anthus trivialis*   14–16 cm

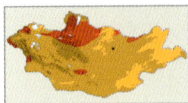

ID A small, boldly marked forest pipit, most likely to be confused with Olive-backed Pipit, which see for distinctions. **Voice** Song is a pleasant, varied series of rapid trills interspersed with repeated short notes, *zit-zit-zit…* sung from tree-top or in parachuting song-flight to ground or base of trees. Call is a drawn-out, hoarse *tseez* or *bizz*; also a sharp alarm, *tseep*, rather like that of Yellow Wagtail. **Habitat** Nests in all types of highland forest, but also occasionally above tree-line as high as 3,500 m. Occurs in other open habitats on passage, including arid steppe and desert. **Behaviour** A conspicuous tree-top singer during breeding season; otherwise rather shy. Nests on ground in shelter of fallen log or similar. Forages on ground with slow, deliberate stride, as well as in trees. **Status** Nominate *A. t. trivialis* is uncommon and local breeding visitor mainly in northern, north-western and central mountain ranges across Mongolia and uncommon passage migrant throughout, late April to early September. Occurrence of *A. t. haringtoni* in W Mongolia is considered doubtful. **Conservation** Considered Near Threatened in Mongolia. Listed in the Mongolian Red Book (2013).

Olive-backed Pipit *Anthus hodgsoni*   14–15 cm

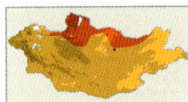

ID Very like Tree Pipit but distinguished mainly by broad white supercilium above and behind eye, made even more prominent by black border above it; and a black-and-white 'ear-patch', which Tree Pipit lacks, except in rare individuals. In good light, dull olive (vs. brown) tone of upperparts can be seen. **Voice** Song, delivered from tree-tops or in display flight, is similar to Tree Pipit's but is variously described as weaker, higher-pitched or drier, with shorter, much more varied phrasing. Call like Tree Pipit's but slightly higher and weaker. **Habitat** Nests in all types of montane forest, and very occasionally in scrub above tree-line. Where it co-occurs with Tree Pipit, may prefer denser, moister forest. **Behaviour** Similar to Tree Pipit. **Status** Fairly common breeding visitor to mountain ranges of north-central Mongolia and common passage migrant, late April to early September (mid-September in the steppe and Gobi). [Alt: Olive Tree Pipit]

Pechora Pipit *Anthus gustavi*   14–15 cm

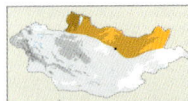

ID Due to its striking white-and-black mantle striping and white wing-bars – as well as its habitat preference and behaviour – this pipit is only likely to be confused with 1st-winter Red-throated Pipit, which see for distinctions. Compared to nominate birds, *A. g. menzbieri* is somewhat darker above and slightly smaller. **Voice** Rarely calls, even when flushed. May give a short dry *tsip* or explosive *chip*, sometimes rapidly repeated. **Habitat** Nests in subarctic tundra and bogs. On migration, favours wetlands with low, dense cover. **Behaviour** Unlike other pipits, very skulking in its habits and difficult to flush. **Status** Rare but regular passage migrant in late April or early May and late August or early September to river valleys in Hövsgöl, northern Hentii and Dornod provinces and Trans-Altai Gobi. Most birds are likely to be nominate *A. g. gustavi* but *A. g. menzbieri* has been recorded in E Mongolia.

Red-throated Pipit *Anthus cervinus*   15 cm

ID Brownish-orange face, throat and breast of adult male (paler and more restricted in adult female) are diagnostic. First-winter birds are readily distinguished from Tree and Olive-backed Pipits by strongly contrasting mantle striping and different face pattern. These are very similar to juv and 1st-winter Pechora Pipits but have finer bill, which is yellowish (vs. pinkish) at base; no buffy coloration on breast; and no dark mark in front of eye. This gives Red-throated a more 'open' expression in contrast to Pechora's 'determined', even 'mean' expression. See also Pechora Pipit for habitat and behavioural differences. **Voice** Flight call is a distinctive, drawn-out, somewhat explosive *pseeeh*. **Habitat** Nests in wet Arctic tundra. On migration, prefers wetlands but also occurs in other open habitats, including arid steppe and desert. **Behaviour** Feeds in open and is less skulking than Pechora or Tree Pipits. Flocks with other pipits on passage. **Status** Uncommon to rare passage migrant in late April to early May and late August to early September, mainly through mountain river valleys.

juvenile

Tree Pipit

adult

trivialis

adult

haringtoni

juvenile

adult

menzbieri

Olive-backed Pipit

adult

gustavi

juvenile

adult

1st-winter

Pechora Pipit

1st-winter

Red-throated Pipit

adult
variant

adult

1st-winter

Common Chaffinch *Fringilla coelebs*
15–16 cm

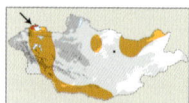

ID Breeding male has unmistakable combination of rosy face and breast, blue-grey hood and bold white wing-bars. Drab female might be mistaken for female House Sparrow, but has pale nape patch and lacks the sparrow's bold streaking on back. Compare also female Brambling. **Voice** Typical song is a 'cheerful', articulate and constantly repeated phrase, usually with decisive wrap-up: *tziv-tziv-tziv-till-till-chiiri-chiiri-iiri-tzii-chewii*. Calls include a sharp *pink*, upward-inflected whistle *hueet* and soft *yup* in flight. **Habitat** Woodland and woodland edge, also locally in scrubby thickets. **Behaviour** Males sing tirelessly from tree perches. Nests are typically fixed snugly into fork in trees or tall shrubs. Forages chiefly for variety of seeds on migration and in winter. Forms loose migratory flocks, often with Bramblings. **Status** Very rare and local breeding visitor and passage migrant in lower Torkholig River valley (Lake Uvs). Also recorded in Great Lakes Depression, Hentii range, Herlen-Ulz River Basin and Gobi on migration; migrants occur mainly in late April and late September. Some individuals winter in mountain forest in Hentii. [Alt: Eurasian Chaffinch]

Brambling *Fringilla montifringilla*
14–16 cm

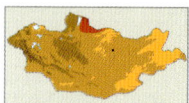

ID Black head and back in combination with orange breast and forewings is diagnostic for breeding male. Non-breeding male is similar but with blotchy black. Female is similar to female Common Chaffinch but has pale orange (vs. buffy-grey) breast and orange (vs. whitish) wing-bar. Both sexes distinguished from Common Chaffinch in flight by white rump. **Voice** Song is a short, harsh, monotone buzz, *tcheeeeeeur*, and rarely, a rattling trill. Calls include a nasal *tsiiv-tsiiv-tsiiv-chiikh* and *chuk-chuk* in flight. **Habitat** Nests in open deciduous, coniferous and mixed forest in highlands. On migration, also visits tree plantations and variety of open habitats where seeds or fruits are plentiful. **Behaviour** Males sing constantly from tree-tops during breeding season. Feeds primarily on seeds on ground during migration, but dried fruit in autumn and wild frozen fruits in late autumn. Migratory flocks of up to 12 birds have been recorded. **Status** Rare breeding visitor locally to parts of Hövsgöl mountain range and Orkhon-Selenge River Basin and fairly common passage migrant in forested habitats throughout, late April to late September.

European Greenfinch *Chloris chloris*
15–16 cm

ID A distinctive, medium-sized greenish-grey finch with heavy pale bill and yellow flashes in wings and tail. Juv is paler with fine streaking above and below. Equally rare Grey-capped Greenfinch is similar in form, but is pinkish-brown overall with contrasting grey head. Records for the two species are from opposite ends of the country. **Voice** Song is highly variable, starting with nasal trill, then rising *tew-tew-tew-tew* interspersed with *tswee* note. Call is a rapid *chwichwichwichwichwi* and single-note *chweeeet* or *tue*. **Habitat** Deciduous woodland and tree plantations in river valleys and towns in west Mongolia. **Behaviour** Breeding pairs nest in planted poplar trees in towns. Gregarious and confiding outside breeding season. Feeds largely on seeds but also takes berries and other fruits on migration. Occurs in flocks outside breeding season, sometimes mixing with other finches and buntings. **Status** Very rare and local breeding visitor in Khovd town and Bulgan village, Khovd province, and rare passage migrant within breeding range, late April to early October. **Taxonomy** Formerly placed in genus *Carduelis*.

Grey-capped Greenfinch *Chloris sinica*
14–16 cm

ID Similar in form to European Greenfinch, but easily distinguished by plumage and distribution. See European Greenfinch for details. **Voice** Song and calls very like European Greenfinch, but somewhat coarser; flight call is a distinctive rapid *tzwi-tzwi-tzwi*. **Habitat** Nests in deciduous and mixed forest and tree plantations. Also frequents dense stands of willow and fruit trees in river valleys during migration. **Behaviour** Gregarious outside breeding season, mixing with other finches on migration and feeding on seeds and fruit both on ground and in trees. **Status** Rare passage migrant in east, where it probably breeds, but nesting unconfirmed; late April to early September. **Taxonomy** Formerly placed in genus *Carduelis*. [Alt: Oriental Greenfinch]

♀ Common Chaffinch ♂ ♀

Brambling ♀ non-breeding ♀ non-breeding ♂ fresh ♂ worn (spring/summer)

European Greenfinch ♀ ♂

♂ Grey-capped Greenfinch

Eurasian Siskin *Spinus spinus* 11–12 cm

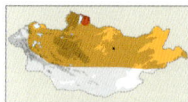

ID A tiny, distinctive finch with very fine, pointed bill, the adult male boldly marked in black and yellow. Female and imm are duller and more heavily streaked. European Greenfinch has yellow flashes in wings and tail like this species, but is much larger and has a notably heavy bill. **Voice** Song is a cacophony of squeaky twittering, with occasional trills and punctuating whistles. Calls include exclamatory squeaking whistles and rattling call, *tititi*. **Habitat** Coniferous and deciduous stands in boreal and mountain forest. Forages for seeds in similar habitats in autumn and winter. **Behaviour** Highly social, forming small, cohesive and sometimes large flocks, especially on migration. Nests in conifers, often high up and usually near end of branch. Very active and restless while feeding. Forages on ground as well as in trees, often with wintering and migratory flocks of Common Redpoll and Twite. **Status** Rare breeding visitor in mature coniferous and mixed forest in Uur River valley (east Hövsgöl mountain range) and possibly nests north to 50 degrees latitude and east to Zelter River valley, Selenge province. Uncommon passage migrant from Great Lakes Depression, Khangai, Hövsgöl and Hentii ranges to east. Wintering flocks have been noted in Hentii mountain range; occurs late April to late September, and sporadically in winter. **Taxonomy** Formerly placed in genus *Carduelis*.

European Goldfinch *Carduelis carduelis* 12–13 cm

ID An unmistakable small finch with a bright red face and extensive yellow flashes on black wings in both sexes. Two distinct plumage-types occur in Mongolia: the grey-headed form *C. c. subulata* is paler and lacks black-and-white head pattern of the typical European form, *C. c. frigoris*. Juv bears slight resemblance to juv European Greenfinch but has whitish, finely streaked head, more pointed bill and much bolder black, white and yellow wing pattern. **Voice** Song is a rapid, twittering *tsswit-witt-witt* and buzzing *zee-zee* notes. Call is a rapid, liquid *tickliit* and harsh *ziiz*. **Habitat** Nests in scattered mature conifers at forest edges and in open woodland with clearings in west Mongolia. Found in open habitats with bushes and trees on migration. **Behaviour** Nest site is typically about 2–5 m above ground, more rarely in tall shrub. A gregarious seedeater. **Status** *C. c. frigoris* is a very rare breeding visitor to upper Bulgan River valley. Migrant birds have been recorded in the Great Lakes Depression, Torkholig River delta (northern Uvs Depression), and possibly Hövsgöl mountain range. *C. c. subulata* breeds from Ölgii town to the Khovd River valley. Migrants occur in the Torkholig River valley, Bulgan River basin, and possibly the Herlen-Ulz River basin and Lake Buir-Khalkh River-Khyangan region. Present late April to mid-September. **Taxonomy** The eastern form, *C. c. subulata*, is sometimes treated as part of a separate species, Grey-headed Goldfinch *C. caniceps*.

Red-fronted Serin *Serinus pusillus* 11–13 cm

ID Black head and heavy streaking plus deep red forehead make adult male unmistakable. Female may be confused with Twite, but is much smaller with dark reddish-brown (vs. tan) face and yellow (vs. pink) rump. Note also status and distribution. **Voice** Song is an extended, disjointed twittering with a whistled quality, *tsrivi-ihihihihihihihi-tsrihi*; call is a metallic twittering, *tshvit, tshvit*. **Habitat** Nests in conifer or mixed forest near tree-line at 1,800–3,000 m. Feeds and migrates through open habitats such as alpine barrens and stony desert. **Behaviour** Gregarious and confiding, especially after breeding season. Sometimes flocks with other ground-haunting seedeaters. **Status** Rare and very local breeding visitor and perhaps rare passage migrant in south-west Mongolia. About a dozen individuals were photographed at Tsagaan Dersnii Bulag oasis, Gobi-Altai province, 21–28 October 2002. One active nest was found in upper reaches of Buduun Khargait River valley, Khovd province, 10–13 June 2013. [Alt: Fire-fronted Serin]

Common Linnet *Linaria cannabina* 12–14 cm

ID Combination of grey head with large red frontal patch, extensively red breast, unstreaked reddish-brown back and whitish wing-patch of adult male is diagnostic. For distinguishing female and imm from Twite, see that species. **Voice** Song is an unremarkable assemblage of short trills, whistles and buzzy chirps. Calls include a characteristic *chewee, chewee*, and in flight *tut-tut, tuttuttutt*, both similar to Twite. **Habitat** Nests on scrubby rocky slopes, and during migration also frequents riverbank copses, deciduous forest edges and parkland, as well as more open habitats. **Behaviour** Similar to other small seedeaters. Breeding and migration in Mongolia poorly known. **Status** Very rare and local breeding visitor in Yolt River valley, upper Khovd River to Bulgan River basin (Mongol-Altai mountain range) and Torkholig River delta (Great Lakes Depression), and very rare passage migrant through breeding areas; presumably late April to mid-September. **Taxonomy** Formerly placed in genus *Carduelis* or *Acanthis*. [Alt: Eurasian Linnet]

Eurasian Siskin

♂

juvenile

adult

juvenile

frigoris

subulata

adult

juvenile

Red-fronted Serin

♂

European Goldfinch

Common Linnet

juvenile

♀

♂

Twite *Linaria flavirostris*   12–14 cm

ID Perhaps best identified by absence of distinctive field marks. An overall brown, heavily streaked finch. Non-breeding birds have yellow bills and may be confused with Common Redpolls, with which Twites often flock in winter, but the latter are tawny-brown (vs. whitish grey) overall, lack a red frontal spot and have a pink (vs. whitish) rump. Female Common Linnet is paler, with grey head, much finer streaking and lacks yellow bill and pink rump. **Voice** Song consists of a few rather harsh, buzzy twanging notes or 'bleats', *twiit, twiit*, reminiscent of the bird's name. Call is a single *chewee*, and a rapid *tzup-tzup-ztup* in flight. **Habitat** Dry, open habitats near springs and streams in rocky gorges, and brushy high mountain slopes; also forest edges, agricultural fields and (rarely) wooded parks in towns and cities out of breeding season. **Behaviour** Usually seen in pairs during breeding season and in small flocks thereafter. Nests in bushes, in tall grasses up to 0.5 m above the ground or on ground among rocks. Flocks with redpolls and other small seed-eating finches in winter. Often gathers at water sources on hot days in dry mountain habitat. **Status** Common resident breeder in high mountains throughout country. Non-breeding, including wintering, birds descend mountain and river valleys to open steppe and Gobi by late October, returning to highlands by late April. **Taxonomy** Formerly placed in genus *Carduelis* or *Acanthis*.

Common Redpoll *Acanthis flammea*   12–14 cm

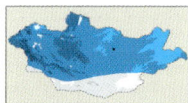

ID The two redpoll species are readily distinguished from Mongolia's other small fringillid finches by their whitish-grey coloration, strong white wing-bars and (especially) red 'poll' above the bill in all adult plumages. More problematic is distinguishing between the two redpolls, for which see the Arctic Redpoll account. **Voice** Song is a rather throaty, dry composite of chirps and running twitters, e.g. *zzrrrrrr*. Call is a frequent *chutt-chutt-chutt-chutt*. **Habitat** Nests in open, mixed birch, conifer and willow woodland typical of boreal and subarctic regions. On migration and in winter seeks out abundant seed sources such as grassy and weedy steppe and birch groves. **Behaviour** Nests in trees at various heights from crown to near ground. A gregarious and irruptive species, sometimes forming flocks of up to 200 birds with Arctic Redpolls, but at other times wholly absent. **Status** Possibly very rare local breeding visitor in Taishir mountain of Gobi-Altai province and northern Hövsgöl mountain range. Sporadic winter visitor from Gobi-Altai, reaching Khalkh River-Khyangan in the east and Baruunkhurai Depression in the west, mainly mid-October to late February.

Arctic Redpoll *Acanthis hornemanni*   12–14 cm

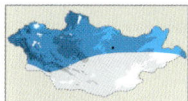

ID Very similar to Common Redpoll, and in some instances it is impossible to distinguish the two species reliably. The most distinctive Arctic Redpolls are very pale (whitish) overall with minimal streaking on flanks; pale rear scapulars (lower back), and (most significantly) little or no black streaking on white rump and undertail-coverts. Arctic also has an even tinier bill than Common. These traits, however, are variable, overlap and are often difficult to see in the field. Fortunately, the two species typically flock together, aiding comparison. **Voice** Song and call very similar to Common Redpoll, though some sources state that they are softer and higher. **Habitat** Nests in Arctic tundra in dwarf willow copses. In autumn and winter, frequents open, dry habitats that are snow-free and have abundant seed sources, such as birches, weeds and grasses. **Behaviour** As Common Redpoll. **Status** Uncommon to rare and irregular winter visitor and late autumn/early spring migrant from the west to Khalkh River-Khyangan, early November to early March. [Alt: Hoary Redpoll]

Twite

♀

♂

♂
fresh

Common Redpoll

juvenile

♀
fresh
(autumn/winter)

♂
fresh
(autumn/winter)

♂
worn
(summer)

♂
breeding

Arctic Redpoll

non-breeding

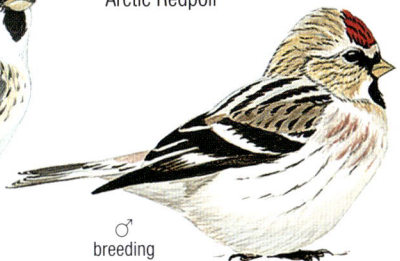

♂
breeding

Plain Mountain Finch *Leucosticte nemoricola*
14–16 cm

ID This aptly named finch resembles adult female House Sparrow in all plumages but lacks prominent pale supercilium and is restricted to high montane barrens. Might be mistaken in non-breeding season for larger, paler Brandt's Mountain Finch with which it co-occurs in western mountains. However, it is darker and more boldly streaked on the back with two prominent pale wing-bars (vs. Brandt's pale wing panel) and shows less white edging in tail. **Voice** Song has been described as 'a sharp twitter', *rick-pi-vitt* or *dui-dip-dip-dip*. Call is soft chirps and twitters – not unlike House Sparrow. **Habitat** Rocky slopes with junipers and other dwarf trees and shrubs in alpine and subalpine zone above 3,000 m. **Behaviour** Nests in rocky cavities and crevices. Sings from top of tall boulders. Often gregarious, gathering at creeks and springs to drink and bathe at midday. Feeds on ground, mainly on seeds of *Artemisia* spp and other plants. Wintering flocks move down to lower mountain slopes in late autumn, returning to breeding elevation in early spring. **Status** Uncommon and local resident breeder and altitudinal migrant in Tsagaan Shuvuut, Kharkhiraa, Turgen and Khan Höhii mountains in Mongol-Altai mountain range.

Brandt's Mountain Finch *Leucosticte brandti*
15–17 cm

ID Breeding adult male is distinctive, with sooty-black bill and nape contrasting with overall silvery-grey plumage with faint pinkish overtones and (often concealed) pink rump. Non-breeding birds are much paler and could be mistaken for Plain Mountain Finch, which see for differences. **Voice** Song is unrecorded. Call is reported to be a sharp and loud *twit-twit*, *tvee-ti-ti*, or *peek-peek* and *churrr*. **Habitat** Nests in treeless alpine meadows and stony barrens on crests of highest mountains from 2,800 up to 5,000 m. Descends to rocky shrublands on lower slopes in winter. **Behaviour** Nests in rock crevices and small clefts in cliffs. Feeds on ground on seeds, buds and fruits of alpine plants as well as invertebrates. Forms small flocks during non-breeding season. **Status** Rare and local resident breeder and altitudinal migrant in massifs of Tavan Bogd and Siilhem (Mongol-Altai) to Gichgene range on border of Gobi-Altai mountains and east to peaks north of Lake Hövsgöl. Winters at lower elevation from late October to early April. [Alt: Black-headed Mountain Finch]

Asian Rosy Finch *Leucosticte arctoa*
16–17 cm

ID Adult plumage is unique – overall sooty-brown to blackish with greyish to buffy nape, heavy streaking below and (most strikingly) a suffusion of rosy pink on upperparts and belly. Female and imm resemble adult male but are duller and greyer. **Voice** Song is a short, explosive two- to three-syllable *chee-ew*. Flight call is a chirp: *piup-piup-piup*. **Habitat** Breeds in rocky tundra in alpine zone near snowline at 2,800–3,000 m. Winters on lower slopes in similar habitat but with more cover. **Behaviour** Nests in crevices and among rocks in mountain barrens with cliffs and large boulders used as singing posts. Forages on ground for seeds and fruits of montane plants as well as invertebrates. Joins mixed flocks of rosefinches and other seedeaters post-breeding. **Status** Uncommon resident breeder and altitudinal migrant. Three subspecies occur in Mongolia: *L. a arctoa* in far western Mongolia including Kharkhiraa and Turgen mountains (Mongol-Altai mountain range) (no breeding records); *L. a. cognata* in Hövsgöl mountain range as a winter visitor; and *L. a. sushkini* in Mongol and Gobi-Altai, Khangai and Hentii mountain ranges. Winters in southern Khangai Plateau and river valleys of upper Herlen-Ulz River Basin and Onon and Balj Rivers. Depending on snow and food availability, descends to lower mountain slopes in November–December and returns to high mountain habitats by February–March.

Mongolian Finch *Bucanetes mongolicus*
11–13 cm

ID Resembles much rarer Desert Finch in size and form, and pink-and-white wing pattern, but distinguishable in all plumages by its pale (vs. black) bill. Adult male also has a suffusion of pink below, which is lacking in Desert Finch. **Voice** Song is repeated *tshwee-tshee-tsvit*. Call is a short *tu-vuiit*. **Habitat** Cliffs, gorges and rocky slopes near springs and creeks in high mountain ranges of arid steppe and desert. **Behaviour** Gregarious, family groups often drinking from open water at midday. Nests on ground or in small cliffside crevice or hollow sheltered by an overhanging rock or bush. Feeds on usual seedeater combination of terrestrial invertebrates while breeding and seeds and fruits during rest of year. Forms post-breeding flocks of up to 80 individuals. **Status** Common but local resident breeder and altitudinal migrant in Mongol-Altai mountain range and Uvs Depression, south through Great Lakes Depression to Baruunkhurai Depression, and east to Valley of the Lakes, the main Khangai range, the Gobi-Altai Range and eastern Gobi. Wintering flocks descend to lower hills in winter. [Alt: Mongolian Trumpeter Finch]

Desert Finch *Rhodospiza obsoleta*
13–15 cm

ID Highly distinctive, unstreaked, sandy coloured finch with jet-black lores and bill, and striking pink, black and white wing pattern (female lacks black lores). For comparison with much more common and widespread Mongolian Finch, see that species. **Voice** Song is a seemingly improvised riff containing a mixture of raspy warbles, trills and squeaks. Calls include a soft *drrrt* and sharp *peek*. **Habitat** Open desert. Feeds in vegetable fields and visits water sources on migration. **Behaviour** Similar to Mongolian Finch. Feeds mainly on seeds of various plants on ground. **Status** Vagrant. One at Khovd town, Khovd province, 8 April 2007. **Taxonomy** Formerly placed in genus *Rhodopechys*.

juvenile

adult

Brandt's Mountain Finch

juvenile

margaritacea

adult

Plain Mountain Finch

♀

arctoa

♂

Asian Rosy Finch

♀

sushkini

♂

♂

♀

♂

Mongolian Finch

juvenile

♂

♀

Desert Finch

♂
worn
(summer)

PLATE 105: FINCHES V – ROSEFINCHES

Common Rosefinch *Carpodacus erythrinus*   15–18 cm

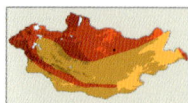

ID Adult male readily distinguished from other species by bright red (not rosy) head, breast and rump, contrasting with white belly and vent. Female and imm from congeners by olive-brown plumage with more diffuse ('smudgier') streaking; notably short, stout, somewhat rounded bill; and two pale wing-bars. **Voice** Song is a whistled three- to four-note *eish-twee-tweeh*. Call is a sharp, upward-inflected *tshweet!* **Habitat** Nests in thickets in taiga and montane forest edges, along lake and river valleys, and in parks. Frequents various open and lightly wooded areas, including steppe and desert regions on migration. **Behaviour** Breeding males sing tirelessly from tops of trees. Nests are constructed in bushes and trees close to ground. Feeds on a wide variety of fruits and seeds. **Status** Common breeding visitor from Mongol-Altai mountain range and Great Lakes Depression in the west, through all major northern mountain ranges and the Orkhon-Selenge River Basin, east to Buir Lake-Khalkh River-Khyangan Region in the far east; and common passage migrant throughout the breeding range south to Gobi, late April to early September. **Taxonomy** Two subspecies, *C. e. erythrinus* and *C. e. grebnitskii*, occur in Mongolia.

Pallas's Rosefinch *Carpodacus roseus*   15–17 cm

ID Adult male readily distinguished by white 'frosting' on otherwise rose-coloured head and throat, and two distinct, pale wing-bars. Head and breast of female and imm are suffused with pink over streaky tan plumage. **Voice** Song is a series of rising and falling notes. Calls include a soft whistled *feee*, metallic *tsuii-i* and bunting-like sharp *tzik*. **Habitat** Nests in coniferous and mixed montane and taiga forest with dense shrubby understorey near tree-line in high mountains. Migrants and wintering birds frequent river floodplains, birch groves and the like at lower altitudes, with some birds remaining in breeding habitat. **Behaviour** Similar to other high-altitude rosefinches. **Status** Uncommon breeding visitor and rare passage migrant from Mongol-Altai mountain range (up to 2,300 m), Great Lakes Depression, and Hövsgöl Mountains, east to Zelter River valley, Selenge province. Very rare wintering species in Khasagt Khairkhan, the main Khangai and Hentii mountain ranges east to Onon, Balj and upper Ulz River valleys. Altitudinal migration early to mid-November and late February to early March.

Red-mantled Rosefinch *Carpodacus rhodochlamys*   16–18 cm

ID The second largest Mongolian rosefinch. Adult male is uniformly dark rose-colour with distinct pale rose supercilium and single wing-bar. Female and imm perhaps best told from smaller alpine rosefinches by size and notably heavy, pale bill. See also Great Rosefinch. **Voice** Song is a high-pitched, rapid series of interspersed squeaks and warbles. Call is a short, rasping, exclamatory *squeenk!* **Habitat** Breeds on open, high-altitude slopes with scattered dwarf juniper and other shrubs as well as in overgrown deciduous thickets in subalpine meadows (1,800–3,000 m). On migration and in winter, frequents open brushy areas, forest edges and cultivation. **Behaviour** Nests low or high in spruces or junipers. Feeding behaviour similar to other rosefinches. **Status** Rare and local resident breeder in Höh Serh massif (Mongol-Altai mountain range); western Khangai Mountains and isolated sites on Högnökhaan Mountain (Khangai range). Rare altitudinal migrant and winter visitor to lower elevations in Hövsgöl, Buyant River valley (Khovd), Khangai mountain range, Orkhon-Selenge River Basin and Hentii mountain range.

Himalayan Beautiful Rosefinch *Carpodacus pulcherrimus*   15 cm

ID Adult male distinguished from other two small, high-altitude Mongolian rosefinches by its pale rosy underparts with darker suffusion of colour on face and upper breast; broad (pink) supercilium; and brown (vs. reddish) crown and back. Female and imm Red-mantled Rosefinch are very similar to Himalayan Beautiful but are significantly larger with paler, larger bills. Compare also Pallas's Rosefinch. **Voice** Song is an undistinguished dry *cheep-cheep chipchipchip*. Call is a sharp *tsinh tsinh*. **Habitat** Breeds among dense thickets of juniper and other dwarf vegetation on rocky slopes at or above tree-line in high mountains. Winters in similar habitat at lower elevations. **Behaviour** Typically nests in low shrubbery. Family groups and post-breeding flocks often visit water sources to drink and bathe. **Status** Uncommon and local resident breeder and altitudinal migrant in Gobi-Altai and western Khangai ranges; north to mountains of the Great Lakes Depression.

Great Rosefinch *Carpodacus rubicilla*   19–20 cm

ID Adult male is unmistakable due to large size; bright, raspberry-red plumage with distinctive whitish speckling on underparts and crown; cheek patch; and heavy, pale bill. Female and imm very like Red-mantled Rosefinch, but slightly larger bill and overall, with generally darker, heavier streaking. **Voice** Song is a clear and variable *chüih, chü-chü-chü twuitwuitwui tsü*. Call is a high and strong *twui* and *chik* when alarmed (Svensson *et al.* 2009). **Habitat** Rocky alpine barrens above 2,000 m. Winters in varied habitats at lower altitudes, e.g. open brushy areas in lake and river valleys, forest edges and plantations. **Behaviour** Nests in low bushes, crevices in cliff faces or hollows at base of boulders. Feeding habits similar to other rosefinches. **Status** Uncommon resident breeder in Mongol-Altai mountain range and Baruunkhurai Depression. Also possibly found on Höh Serh Mountain and Arts Bogd in Gobi-Altai range; and Khan Höhii Mountain (western Khangai range). Some birds winter in breeding areas as well as overgrown thickets in Valley of the Lakes and Gobi. Descends to lowlands by late November and returns to breeding sites by mid-April.

♂

♂

♀

♂
1st-summer

Common Rosefinch

erythrinus

Pallas's Rosefinch

♀

♂

Red-mantled Rosefinch

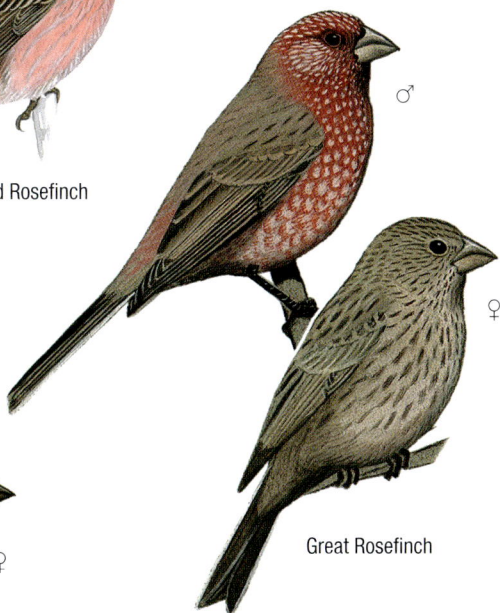

♂

♀

Great Rosefinch

♂

♀

Himalayan Beautiful Rosefinch

Long-tailed Rosefinch *Carpodacus sibiricus*   15–18 cm

ID Adult male unmistakable with combination of pale pink suffusion on head and breast; extensive white in wings; pale bill; and notably long tail. In good light has a 'frosted' look (cf. Pallas's Rosefinch). Female and imm somewhat resemble female redpolls, but lack red frontal spot and have more white in wing and much longer tail. **Voice** Song is a 'cheerful', whistled five- to seven-note, *swechoo-cheeru cheery* or similar. Call is an abrupt, sharp, single or multiple whistled notes: *fiu fiu fiu*. **Habitat** Nests in young, dense thickets of willow and other deciduous trees along riverbanks and edges of taiga and montane forest. Winters in similar brushy habitats at lower elevations. Also occurs in wooded parks. **Behaviour** Nests in birches and poplars, etc., 1–2 m above ground. Feeds mainly on seeds, buds and fruits in trees and on ground and takes terrestrial invertebrates during breeding season. **Status** Uncommon resident breeder and altitudinal migrant from Mongol-Altai mountain range to Uvs Lake Depression, through Khangai, Hövsgöl and Hentii mountain ranges and Orkhon-Selenge River Basin east to Buir Lake-Khalkh River-Khyangan Region in the far east. Descends to lower-altitude hills and brushy cultivated areas in winter. **Taxonomy** Formerly placed in genus *Uragus*.

Pine Grosbeak *Pinicola enucleator*   19–22 cm

ID A distinctive, large, full-bodied arboreal finch with stubby bill and relatively long tail. Superficially resembles Two-barred Crossbill as males of both species have largely pink plumage and prominent white wing-bars. However, Pine Grosbeak is much larger, has white tertial edges as well as wing-bars and lacks crossed mandibles. Female Two-barred Crossbill is heavily streaked vs. uniform grey plumage of female grosbeak. Largest rosefinches lack white wing-bars and are ground (vs. tree) feeders. **Voice** Song is a pleasant musical warble, e.g. *piliup-piliup-pilidipu-piliup*. Call is clear and loud, *pluiu* or *piu-piu*. **Habitat** Breeds in conifer forest of montane taiga, often with dense shrub layer. In winter, descends to lower elevations where it seeks out habitats with fruit- and seed-bearing trees and shrubs, including parks and gardens. **Behaviour** Nests next to trunk, deep in conifer forest, especially spruce. An irruptive species, appearing in large flocks in some winters while absent in others. Shy when breeding but often very approachable during winter. **Status** Uncommon resident breeder from Mongol-Altai to Khangai, Hövsgöl and Hentii mountain ranges. Wintering flocks typically move to lower elevations in November and return to breeding areas by late February–early March.

Red Crossbill *Loxia curvirostra*   15–17 cm

ID Only Two-barred Crossbill shares this species' distinctive crossed mandibles, and it is easily distinguished vs. both sexes by its bold wing-bars. Superficially resembles Common Rosefinch but has red and grey (vs. white) belly in addition to unusual bill. Other small rosefinches are ground-feeders (vs. tree-feeders) and are unlikely to occur together with crossbills. **Voice** Song is an apparently random series of chirps, short trilled notes and chatters. Call is a hard, sharp, usually repeated *kiupp-kiupp-kiupp*. **Habitat** Breeds and winters mainly in coniferous forest, but also mixed forest from high-altitude mountain taiga forest down to middle elevations. **Behaviour** An arboreal, flocking seed-eater, its unusual bill is adapted for extracting seeds from cones of pines and other conifers; also feeds on fruit. Nest is sited next to main trunk of conifer trees at 1.8–12 m. An irruptive species that, depending on food availability, snow cover and air temperature, moves in varying numbers down to river valleys, forest edges and cultivated trees in towns and cities. **Status** Uncommon resident breeder and altitudinal migrant from Mongol-Altai mountain range to eastern Onon and Balj River valleys. Disperses to lower altitudes from November, returning to breeding areas from late February to early March. [Alt: Common Crossbill]

Two-barred Crossbill *Loxia leucoptera*   14–16 cm

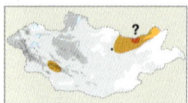

ID Crossed mandibles and plumage characters, especially bold white wing-bars are highly distinctive. Likely to be confused only with much larger Pine Grosbeak, which see for differences. **Voice** Song is a variable, somewhat musical series of chirps, twitters and trills; call a rapid series of three or more dry notes: *pip-pip-pip...pip-pip-pip-pip...* ('lighter' than Red Crossbill call) with occasional nasal bleats. **Habitat** Breeds in forests of Siberian Larch (*Larix sibirica*) and pines. Disperses into deciduous forest and open habitats in river valleys and steppe in seasonal movements. **Behaviour** Generally similar to Red Crossbill, nesting 2–10 m above ground in conifers and prone to irregular irruptive movements outside of breeding areas. **Status** Very rare resident breeder in upper Hentii mountain range and forested areas on Khan Taishir Mountain (Gobi-Altai range). The species has also been recorded at Lake Chukh, May 2009 and Ulz River, 13 July 2009, both in Dornod province; in conifer forest at Bogd Khaan Mountain, 28 January 2007; at Gachuurt, near Ulaanbaatar, every winter; and at Jargalantkhairkhan Mountain, Khovd province, 25 June 2012. [Alt: White-winged Crossbill]

♂ breeding

♀ non-breeding

♂ non-breeding

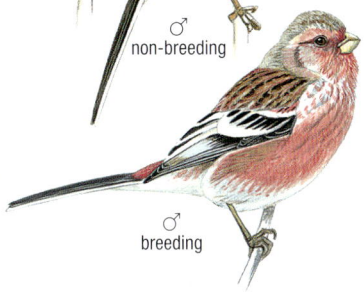

♂ breeding

Long-tailed Rosefinch

♂

♀

Pine Grosbeak

♂

♀

juvenile

Red Crossbill

♂

♀

juvenile

Two-barred Crossbill

Eurasian Bullfinch *Pyrrhula pyrrhula*   15–17 cm

ID Adult male of nominate race is unmistakable with its combination of salmon-pink underparts; grey back; black cap, wings and tail; single broad white wing-bar; white rump; and stubby bill (creating a unique 'bull-headed' look). Female hardly less distinctive but lacks pink breast. Juv is duller and lacks black cap. In *P. p. cineracea* the male is pale grey (vs. pink) below and the female brown above as well as below. **Voice** Song is a deliberate, seemingly random series of squeaks, whistles and especially a clear, melancholy whistle that dominates. Call is a melancholy whistle, *phe-ew*. **Habitat** Breeds in mature mixed and coniferous montane forest with shrubby undergrowth. In winter, frequents more open habitats near forest edges as well as village and city parks where seed- and fruit-bearing trees are dominant. **Behaviour** Nests in deciduous trees and tall bushes 1.2–2.1 m above ground. Eats seeds, fruits and buds in vegetation or on ground. Wintering birds may form mixed feeding flocks with Hawfinches and Bohemian Waxwings. **Status** Rare and local resident breeder in main range of Khangai and Hentii Mountains. Winters within breeding areas and from Tes River valley and northern Khangai mountain and Hövsgöl mountain ranges east to Mongol Daguur Steppe and Lake Buir-Khalkh River-Khyangan Region. Post-breeding flocks descend to lower elevations from late November to early February. **Taxonomy** Two races occur: nominate *P. p. pyrrhula* and *P. p. cineracea*. The latter is sometimes treated as a separate species (Baikal Bullfinch).

Chinese Grosbeak *Eophona migratoria*   16–19 cm

ID Resembles Hawfinch in size and form but readily distinguished by its black (vs. tawny-brown) head and yellow (vs. blue-grey) bill (adult male) or grey head and grey (vs. tan) underparts (female and imm). Distinctive black hood in adult male, dark grey in adult female. **Voice** Song is an articulate loud, clear whistle: *chee chee churee kirichu*, recalling Common Rosefinch. Call is a short, sharp single or multiple squeak *tseek* or drier *tsek*. **Habitat** Young deciduous forest and dense brush in hills and river valleys. **Behaviour** Nests in deciduous trees, including elm, 1.5–2 m above ground. Feeding habits similar to Hawfinch. **Status** Little known in Mongolia; possibly a rare and local but regular breeding visitor and passage migrant in far east. Apparent but unconfirmed breeding in Nömrög River valley and may migrate through Ih Khyangan mountain range (Buir Lake-Khalkh River-Khyangan Region). Also a single bird photographed in planted poplars in Bor-Öndör town, Hentii province, 8 June 2009. Possibly present late April to mid-September. [Alt: Yellow-billed Grosbeak]

Hawfinch *Coccothraustes coccothraustes*   16–18 cm

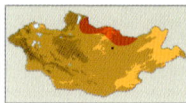

ID Chunky, short-tailed finch with massive bill and white flashes in wings and tail. Likely to be confused only with very rare and local Chinese Grosbeak, which has a black (vs. tawny-brown) head and yellow (vs. blue-grey) bill (adult male), or grey head and grey (vs. tan) underparts (female and imm). Sexes and ages essentially similar. **Voice** Call is a sharp, high, lispy squeak *tziik*. Quiet, seldom heard song is rather vague stringing together of call-notes and other lisping sounds. **Habitat** Nests in deciduous and mixed forests, patchy woodland and mixed trees along river courses. Occurs in arid steppe and desert on migration, and winters in urban parks, gardens and other habitats where its preferred food is available. **Behaviour** Rather shy and quiet and therefore may be overlooked despite its size and gregarious habits. Nests in deciduous trees 1–5 m above ground. Feeds primarily on mast of elm and other trees as well as kernels of cherries and other fruits, which it crushes with its massive, powerful bill. Post-breeding Hawfinches occur in pairs or small flocks of 5–16 individuals, sometimes joining mixed flocks of other finches in winter. **Status** Rare and local resident breeder and partial migrant from eastern Hövsgöl mountain range across Eg and Selenge River basins to Hentii range. On migration, occurs unpredictably throughout, including oases in Gobi. Breeding populations are present late April to late September.

juvenile

pyrrhula

♀

cineracea

♂

♂

Eurasian Bullfinch

♂
breeding

♀

♂
non-breeding

Chinese Grosbeak

♂
juvenile

♂
breeding

Hawfinch

Yellowhammer *Emberiza citrinella* 16–17 cm

ID Combination of yellow head and reddish-brown breast and rump of breeding male is diagnostic. Adult female and imm distinguished from imm and adult female Pine Bunting by yellowish (vs. whitish or buff) tinge below (though this can be hard to see). Hybrids with Pine Bunting can show a variety of intermediate characters. **Voice** Often sings, *zi-zi-zi-zi-zi-zi-zrii-zreeee*, from tops of trees and bushes. Call is a metallic *tsink*, or *zee* and *zit-zit-zit-zit*. **Habitat** Open, thin forest thickets and tall bushes, willow trees in forest steppe and valleys of lakes and rivers. Found in open, brushy habitats and forest edge on migration. **Behaviour** Feeds on ground, foraging for seeds and invertebrates (breeding season) and in trees (winter). Nest placed on or near ground, hidden in grass or at edge of some taller shrubbery. Forms small flocks with mixed groups of other seed-eating birds on migration. **Status** Very rare and local breeding visitor to Orkhon River valley and rare passage migrant through breeding area and Hövsgöl and Hentii mountain ranges, mid to late April to late September.

Pine Bunting *Emberiza leucocephalos* 16–18 cm

ID Eye-catching black, white and chestnut head pattern of breeding male most likely to be confused with that of Meadow Bunting, but note white crown and chestnut throat as well as habitat. Adult female and imm lack yellowish tinge below of Yellowhammer and are much more heavily streaked below than Meadow Bunting in these plumages. **Voice** Song and calls virtually identical to Yellowhammer's with which it sometimes hybridises. Like that species, it often sings and calls from the tops of trees. **Habitat** Forest margins and clearings, mostly coniferous but also deciduous with shrubs, or in open country with young deciduous trees and bushes near water. On migration, favours open areas with tall grasses and bushes from forest to desert. **Behaviour** Nests on ground in thick grass, dense brush and young tree stands. Feeds on terrestrial arthropods in breeding season; forages on ground for grass seeds on migration. Pine Bunting x Yellowhammer hybrids have not been recorded in Mongolia. **Status** Common breeding visitor in all forested mountain ranges and uncommon passage migrant, late April to late August (mid-September in the steppe).

Rock Bunting *Emberiza cia* 16 cm

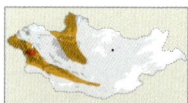

ID Closely resembles Godlewski's Bunting in all plumages, but note black (vs. chestnut) crown and eye-stripes. Meadow Bunting lacks contrasting central crown-stripe. **Voice** Song is a longish, high-pitched, rather melodic jumble of disparate syllables, not unlike Common Reed Bunting. Call is a sharp *tsee* or *zie*, repeated in alarm; also short *tsiip*. **Habitat** Rocky slopes with shrubby thickets and scattered deciduous and larch trees; also high mountain gorges with tall grass and alpine meadows below tree-line. On migration, favours open, dry habitats and brushy forest edges in mountains and river valleys. **Behaviour** Breeding males sing (rather infrequently) from tops of rocks and cliff faces. Builds cup of bark strips, mosses and grass in cavity among rocks or on stony slope, rarely in low bushes among stones. Eats and feeds nestlings terrestrial arthropods in breeding season. Forages on ground and in vegetation for seeds and berries on migration. **Status** Very rare breeding visitor in high mountains of Mongol-Altai mountain range up to 2,700 m. Visits Gobi-Altai mountain range, Tes River valley, Khangai and Hövsgöl ranges and Bulgan River valley in west on migration. Late April to late August (early September in the steppe).

Godlewski's Bunting *Emberiza godlewskii* 17–18 cm

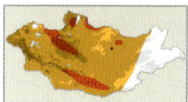

ID Very similar to Rock Bunting (with which it was formerly considered to be conspecific), but adults readily distinguished from that species by chestnut (vs. black) crown and eye-stripes. Juvs of two species may be indistinguishable in the field. Somewhat similar Meadow Bunting lacks contrasting central crown-stripe; compare also face pattern. **Voice** Song and calls are very like those of Rock Bunting. **Habitat** Gorges, cliffs and rocky slopes in high mountains (up to 2,700 m) with scattered tall bushes, often near streams. Post-breeding birds found in all types of dry, open, rocky montane habitats. **Behaviour** Very similar to Rock Bunting. Wintering individuals descend from higher elevations to lowlands and mountain valleys. **Status** Uncommon breeding visitor and winter resident at lower elevations from Mongol-Altai mountains to southern Gobi-Altai mountain range; also, Khangai, Hövsgöl and Hentii ranges.

1st-winter

♂ non-breeding

♀

♂ breeding

Yellowhammer

1st-winter

♂ non-breeding

♀ breeding

♂ breeding

Pine Bunting

♀

♀ 1st-winter

♂

♂

Rock Bunting

Godlewski's Bunting

Meadow Bunting *Emberiza cioides* 16–17 cm

ID Very similar to (vagrant) Jankowski's Bunting, which see for distinctions. Juv and non-breeding individuals might be mistaken for those plumages of Rock and Godlewski's Buntings but always lack the pale central crown-stripe of those species. Breeding male Pine Bunting has white (vs. chestnut) crown. Bold facial markings of adult Rustic and Tristram's Buntings are black (vs. chestnut). **Voice** Song is very similar to Rock and Godlewski's Bunting – a cheerful jumble of disparate syllables – but shorter and higher-pitched. Call is a sharp *zit-zit-zit*. **Habitat** Dry, rocky slopes with scattered bushes at edges of forests in lower mountains (up to 2,200 m). **Behaviour** Nesting and feeding habits very like Godlewski's Bunting. Breeding males sing from tops of bushes and rocks near nesting site. After breeding season, descends from higher elevations to open steppe and valleys. **Status** Common breeding resident from Mongol-Altai mountains through main mountain ranges to highlands of Mongolian far east; may also breed in desert steppe of the Gobi-Altai. Common winter resident at lower elevations within breeding range.

Jankowski's Bunting *Emberiza jankowskii* 16–17 cm

ID Very similar to Meadow Bunting, but cheek patch blackish, not chestnut; underparts whitish, not rufous and with diagnostic dark belly smudge; and narrow pale wing-bars. **Voice** Song is reported as a short *hsuii-dzia-dzia-djeee*, sung from tops of bushes and tall grass. Call is a sharp *tsiit* or *tsiit-tsiit*. **Habitat** Dry, open habitats on sandy soils with grasses and bushy cover; also poplar and conifer plantations. **Behaviour** Similar to other buntings. **Status** Vagrant or possibly very rare passage migrant and winter visitor in Khalkh and Nömrög River valleys and Lake Buir, Dornod province. A single sight record in Petro-Matad oilfield, Dornod province, 19 November 2013, and uncertain records in southern Alashani and eastern Gobi in winter. **Conservation** Globally Endangered due to limited range, small and declining population (< 500 pairs) and continued conversion of its habitat to agriculture. [Alt: Rufous-backed Bunting]

Chestnut-eared Bunting *Emberiza fucata* 16 cm

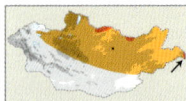

ID Combination of chestnut cheek and carpal patches, white throat, black 'necklace' and whitish belly of adult is highly distinctive. Little Bunting has wholly chestnut face and brow, but is much smaller and lacks black necklace. **Voice** Song is a rather short, rapid, abrupt twitter difficult to render in syllables. Call is a strong and explosive *pzik* or *zii*. **Habitat** Scrubby hillsides near open meadows, dry shrublands and riverside thickets in forested steppe and highland forest edges. Uses a variety of open and lightly forested habitats on migration. **Behaviour** Similar to other buntings. Forms small flocks of 5–10 individuals during migration. **Status** Very rare local breeding visitor to Khangai and Hentii mountain ranges and Nömrög and Khalkh River valleys in far east. Uncommon passage migrant throughout country except for high mountains and desert. Generally present late April to early September.

Little Bunting *Emberiza pusilla* 12–14 cm

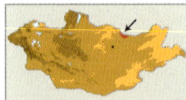

ID Its small size, chestnut face and black streaking on white underparts make this common migrant bunting all but unmistakable. However, see Chestnut-eared Bunting for comparison. **Voice** Song is a high-pitched but melodic, rapid rolling *trutrutrutru srisrisri tuy sivi-sivi chu si*. Call is a sharp, short, clicking *zick*. **Habitat** Nests in birch and willow scrub near swamps and clearings in taiga forest. Occurs in all habitats from boreal forest to Gobi Desert during both spring and autumn migrations. **Behaviour** Feeds on terrestrial invertebrates and plant matter, as well as insects in bark of deciduous and conifer trees on migration. Forms flocks of up to 80 on migration. **Status** Common passage migrant across the country, late April to early May and late August to mid-September. Possibly breeds in mountain taiga forest in Minj River of Hentii mountain range and Hövsgöl range.

Meadow Bunting

♂ 1st-winter

♂

♀

Jankowski's Bunting

♂ non-breeding

♂ breeding

♀ breeding

Chestnut-eared Bunting

♂ 1st-winter

non-breeding

♂ breeding

Little Bunting

Common Reed Bunting *Emberiza schoeniclus*   13–15 cm

ID Similar in all plumages to Pallas's Reed Bunting, but larger and more heavily built with generally darker, more rufous (vs. paler and greyer) coloration above. See Pallas's Reed Bunting and Japanese Reed Bunting for detailed differences. **Voice** Song is a brief outburst of sharp, rolling notes *srip, srip, sria srisrisirr*; this varies in tempo with different populations. Call is a dull, sparrow-like *tschrip* or *tsiu*. **Habitat** Nests in rank vegetation and scrub near water in lake and river valleys, as well as in reedbeds. Occurs in diverse habitats, including open dry steppe, as well as wetlands on migration. **Behaviour** Breeding males sing from tops of reeds or similar perches. Nest is built on the ground, often in grass or rush clumps, or in low bushes. Follows typical bunting diet of invertebrates and seeds. **Status** Common breeding visitor in appropriate habitat across northern Mongolia from Mongol-Altai mountain range and Great Lakes Depression through central mountain ranges and Mongol Daguur Steppe to far east. Common passage migrant throughout the rest of country, including Gobi. Migrants occur mainly late April to early September, in flocks of up to 20. However, large-billed race, *E. s. pyrrhuloides*, breeding in western Mongolia has a heavier bill than other subspecies, allowing it to crush reed stems in search of dormant invertebrates and so enabling birds to overwinter in Mongolia. Other races in Mongolia are *E. s. passerina* (NW Mongolia), *E. s. pyrrhulina* (NE Mongolia) and *E. s. parvirostris* (W Mongolia). [Alt: Reed Bunting]

Pallas's Reed Bunting *Emberiza pallasi*   13–14 cm

ID Very similar to Common Reed Bunting, but smaller and slighter (size of Little Bunting). In addition, breeding male distinguished from Common Reed by greyer (vs. rufous-tinged) upperparts, showing off bold black streaking and whitish wing-bars; also has buff (vs. white) nape; and black bib does not extend below upper breast. Female, non-breeding male and imm are brown, but compared with corresponding plumages of Common Reed, are paler with more contrasting striping above and much lighter (vs. bold, dark) flank streaking; also has notably finer bill with pink (vs. dark grey) lower mandible. See also description for Japanese Reed Bunting. **Voice** Song is a distinctive monotone series of rasping *srih-srih-srih-srih-srih* notes given from tops of reeds or other vegetation. Call is a short and sharp *tsi tsi*. **Habitat** Nests in wide variety of habitats, including dense willow, birch and alder thickets on lake and riverbanks and boggy edges of taiga forest as well as reedbeds and clumps of needle grass (*Achnatherum splendens*) in dry steppe. Uses many, mostly open, habitat types during migration. **Behaviour** Similar to other buntings; nests on ground. **Status** Common breeding visitor and passage migrant throughout country. Absent as breeding species in the most arid southern reaches of Gobi. Late April to late August (early September in the Gobi). Two races occur: *E. p. pallasi* (W Mongolia) and *E. p. lydiae*. [Alt: Pallas's Bunting]

Japanese Reed Bunting *Emberiza yessoensis*   14–15 cm

ID Resembles both Common Reed and Pallas's Reed Buntings but is smaller and slighter than former with rufous (vs. white) nape in breeding males and much more rufous above than latter. Unlike breeding male of both Common Reed and Pallas's Reed Buntings, Japanese Reed has complete black hood without white moustachial slash of Common Reed and Pallas's Reed. Non-breeding plumages have rufous cast to plumage (especially at nape), vs. tawny (Pallas's Reed) or dark brown with grey nape (Common Reed). See Common Reed and Pallas's Reed Buntings for additional plumage differences. **Voice** Song is a brief, rapid twittering, not unlike Meadow Bunting: *chui-tzui-chirin*. Call a short *tsick* or *chu chu chi*. **Habitat** Nests in open marshland and wetland fringes with low reeds, grasses and sedges to height of 30–45 cm, as well as lake and riverbank shrubbery. **Behaviour** Nests on or very close to ground. When flushed it usually flies some distance away. Feeding habits like those of other buntings. **Status** Rare and local breeding visitor to valley of Khalkh River and Lake Tashgain Tavan, Dornod province, and possibly rare passage migrant to Mongolian far east. **Conservation** Considered Near Threatened globally and in Mongolia, due to small and declining population within a limited distribution. Listed in the Mongolian Red Book (2013). [Alt: Ochre-rumped Bunting]

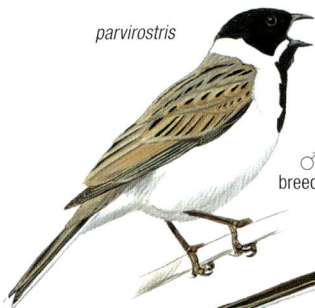

parvirostris

♂
breeding

Common Reed Bunting

pyrrhulina

pyrrhuloides

♂
breeding

♂
breeding

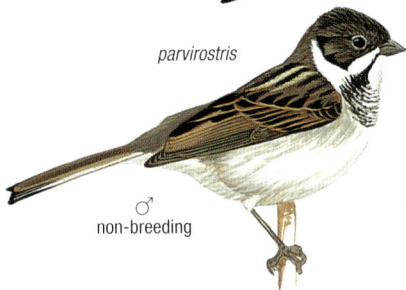

parvirostris

♂
non-breeding

parvirostris

♀
breeding

lydiae

♂
breeding

pallasi

♂
breeding

Pallas's Reed Bunting

♀

♀
1st-winter

♂
non-breeding

continentalis

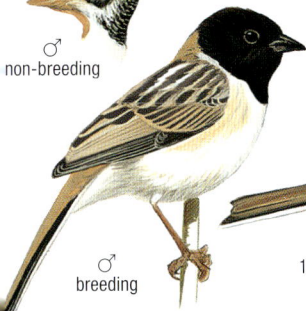

♂
breeding

♀
1st-winter

Japanese Reed Bunting

♀

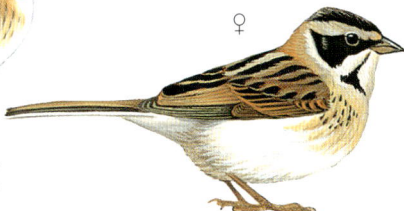

Yellow-throated Bunting *Emberiza elegans*   15–16 cm

ID Bold black, white and yellow head pattern, prominent crest and black chest crescent render breeding male unmistakable. Female is much duller, but crest and head pattern still diagnostic. **Voice** Song is a long, monotone warbling twitter, e.g. *chep-chip-chip-chiririee-chip-chee*. Call is a sharp *tzip* (alarm); also *tzi tzi-tzi*. **Habitat** Prefers dry hillside clearings of deciduous and mixed forest, including plantations. **Behaviour** Nesting and feeding behaviour similar to other buntings. **Status** Vagrant or possibly very rare spring and autumn migrant in Mongolian far east. Known from single sight record of six individuals in garden in Choibalsan town, Dornod province, 28 September 1994. [Alt: Elegant Bunting]

Yellow-browed Bunting *Emberiza chrysophrys*   14–15 cm

ID Might be confused with (rare) Tristram's Bunting, but yellow brow is distinctive in adults. Also, adult male Tristram's has black (vs. white) throat and adult Tristram's of both sexes have rufous-streaked (vs. whitish) flanks. Compare also Rustic Bunting. **Voice** Song is a clear, short and rolling *hwii tii-tii tu-tu-tu*. Call is a sharp, thin, metallic *tsip*. **Habitat** Mixed forest with low conifer trees and edges of coniferous forest along major rivers and their tributaries in eastern Mongolia. **Behaviour** Makes a messy, loose nest of fine grasses and animal hair placed 1–2 m above ground in trees; rarely on ground. Feeding behaviour similar to other buntings. **Status** Rare passage migrant through Herlen-Ulz River Basin and Buir Lake-Khalkh River-Khyangan Region in Mongolian far east; may nest in the Khalkh River valley; possibly present late April to early September.

Tristram's Bunting *Emberiza tristrami*   15–16 cm

ID Adult male of this rare bunting is most likely to be confused with adult male Rustic Bunting but has black (vs. white) throat. Adult female is distinguished from Rustic by latter's bright rufous nape and breast streaking. Adult female is also similar to Yellow-browed Bunting, which in addition to its distinctive yellow brow has prominent whitish wing-bars and lacks rufous wash and streaks on flanks. **Voice** Song is a variable, *ihsiwi-swi-chewi-tzi-tzi*. Call is an explosive *tzik*. **Habitat** Breeds in coniferous taiga and mixed forest with dense undergrowth. Migrants occur in open and shrubby grassland. **Behaviour** Nests on ground amid thick ground cover, but forages in open habitats on migration, feeding seasonally on invertebrates and seeds. **Status** Vagrant or possibly very rare passage migrant to far eastern Mongolia. Presently known from just two records: a sight record of one in Nömrög River valley, Dornod province, in June-July, 1994; and one photographed near Menen military checkpoint, Dornod province, 31 May 2008.

Rustic Bunting *Emberiza rustica*   13–15 cm

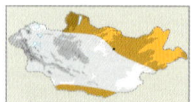

ID Breeding adults unmistakable with combination of black-and-white head pattern; bright rufous nape, breast and flank streaking; unmarked white belly; and small crest. Non-breeding and imm birds may be confused with Yellow-browed and Tristram's Buntings and Lapland Longspur, which see for differences. **Voice** Song is a melodic, energetic, yet slightly melancholy *duu-dili-duudo-diluu-delu* or similar. Alarm call is a sharp, thin *zit* or *zip*. **Habitat** Nests in boreal coniferous and mixed forest edges, and birch and willow groves along riverbanks; frequents these as well as more open habitats during migration. **Behaviour** Nesting and feeding behaviour similar to other buntings. **Status** Uncommon passage migrant across Hövsgöl and Hentii mountain ranges, Eastern Mongolian Plain and Buir Lake-Khalkh River-Khyangan Region. Migrates through oases and bushy mountain slopes in Gobi in small numbers. Late April to early May and late August to early September.

♂

♀

Yellow-throated Bunting

♀ breeding

♂ non-breeding

♂ breeding

Yellow-browed Bunting

♀ breeding

♂ non-breeding

♂ breeding

Tristram's Bunting

♂

♂ breeding

♀ breeding

Rustic Bunting

1st-winter

Black-faced Bunting *Emberiza spodocephala*   13–14 cm

ID Breeding male has unique combination of ashy-grey head (with black face), nape, neck and breast; pale yellow underparts with finely streaked flanks; and warm brown back streaked with black. Plumages of female and imm resemble those of Chestnut and Yellow-breasted Buntings, but are greyer (less yellowish on head and underparts) with a warm brown back (vs. olive-toned). **Voice** Song is a choppy, unmusical mix of trills, lisps and chirps, e.g. *hsin-si-tsiz-tsizrri-tsin-chi*). Call is a sharp *tsii*. **Habitat** Nests in wide variety of broadleaf habitats including young deciduous forest with dense understorey, taiga birch groves and riverbank copses, often near open meadows. **Behaviour** Rather shy. Often sings from hidden perch in dense vegetation. Constantly flicks tail, revealing white outertail-feathers. **Status** Common breeding visitor in Khangai and Hentii ranges and far east. Common passage migrant from Great Lakes Depression east to Buir Lake-Khalkh River-Khyangan Region and south through Gobi, late April to early September (mid-September in the Gobi).

Yellow-breasted Bunting *Emberiza aureola*   14–15 cm

ID Breeding male is unmistakable with striking combination of dark chestnut upperparts, black face and yellow underparts. Adult female distinguished from Chestnut Bunting by darker, more clearly defined striping on head, back and flanks; clear light yellow underparts; and prominent white wing-bars. Juv separated from Chestnut Bunting by stronger head pattern; white wing-bars and bold black striping. **Voice** Individual songs highly variable but typically consist of loud, clear, often paired notes that move up and down in tone and pitch, e.g. *tru-tru trau-trau triihe* with last note long and falling. Call is a soft *tzip tzip*. **Habitat** Scattered willow or young birch thickets in wet meadows and along lake shores and riverbanks; also edges of young forest and forest clearings in mountain taiga and forested steppe. Frequents urban parks with trees on migration. **Behaviour** Nests low in bushes or among tall coarse herbage. **Status** Formerly fairly common, but now rare breeding visitor and uncommon passage migrant in suitable habitats from Mongol-Altai mountain range to east Herlen-Ulz River Basin and up to Buir Lake-Khalkh River-Khyangan Region; mid-April to mid-September. **Conservation** Globally Endangered and Near Threatened in Mongolia. Listed in the Mongolian Red Book (2013).

Chestnut Bunting *Emberiza rutila*   13–14 cm

ID Bright, unstreaked chestnut upperparts and nearly clear yellow underparts of breeding male impossible to confuse with any other bunting (Red-headed Bunting is yellowish-green above and streaked). Adult female and juv are most like Yellow-breasted Bunting. **Voice** Song is a loud, 'declarative' phrase, often beginning with three pure, short whistles followed by rapid 'scratchy' sequence and falling or abrupt 'wrap-up', e.g. *cheer cheer chichichi chui chui chuchuchuu*. Call is a short, sharp clicking *tzit*, sometimes repeated. **Habitat** Nests in coniferous forest with shrubby ground cover and birch or alder thickets in taiga regions. Found at grassland edges, brushy areas and reedbeds in steppe and other open habitats on migration. **Behaviour** Breeding males often sing unobtrusively from concealed perch. Feeding habits as other buntings. **Status** Very rare breeding visitor to Eg and Uur River valleys, Buteel and Khantai mountain and Hövsgöl mountain range, and rare passage migrant from Hövsgöl through central Mongolia and Hentii mountains to far east, late April to early September.

Ortolan Bunting *Emberiza hortulana*   16–17 cm

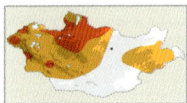

ID Very like Grey-necked Bunting, which see for differences. **Voice** Song is a variable *dsii-dsii-dsii-dsii-huu-huu* or *zree-zree-zree-zuuui-ti-ti-tiu-tiu-tiuuu-u*, similar to Grey-necked. Call is a metallic *tziie*. **Habitat** Breeds in dry scrubby areas in rocky hills and at edge of coniferous forest in high mountains and associated river valleys. On migration, visits wide variety of open scrubby habitats. **Behaviour** Nesting and feeding behaviour similar to Grey-necked Bunting. **Status** Uncommon breeding visitor and passage migrant from western mountain conifer forest region (Khovd and Bulgan Rivers) east to Khan Höhii, across Khangai, Hövsgöl and Hentii mountain range and Orkhon-Selenge River Basin. In addition to breeding range, migrants occur in Middle Khalkh Steppe, eastern Mongolia and northern Gobi; late April to early September.

Grey-necked Bunting *Emberiza buchanani*   16 cm

ID Very similar to Ortolan Bunting, best distinguished by rufous (vs. grey) breast, lacking sharp demarcation between grey and rufous in Ortolan. Throat and malar stripe are cream-coloured (vs. yellow in Ortolan) and the head is clear grey (vs. greenish tone in Ortolan). 1st-winter and juv Ortolan have broader streaks on mantle and breast than Grey-necked, but distinguishing imm birds likely to be challenging. **Voice** Song is an emphatic, ringing *tsui-tsui-tsui-tsuiuu-tsurii*. Call is a low *tseep*. **Habitat** Dry, rocky slopes in mountain valleys with sparse bushes in high mountains. **Behaviour** Breeding males sing from top of rocks and cliff faces. Ground nests consist of dried grass stems lined with fine grass and horse hairs. During hot summer days, birds descend to streams to drink and bathe. **Status** Uncommon breeding visitor and passage migrant from mountain massif in Baruunkhurai Depression east through eastern end of Gobi-Altai range, mountains surrounding Valley of the Lakes, and Southern Khangai Plateau and mountains; late April to early September.

sordida ♂

spodocephala ♂

♀

Black-faced Bunting

aureola

ornata

♂ breeding

breeding ♂

♀

Yellow-breasted Bunting

juvenile

♂

♀

1st-summer ♂

1st-winter ♀

Chestnut Bunting

1st-winter

1st-winter

Ortolan Bunting ♂

Grey-necked Bunting ♂

Black-headed Bunting *Emberiza melanocephala*   16–17 cm

ID Combination of black head, unstreaked chestnut upperparts and clear yellow underparts make the breeding male of this rare bunting unmistakable. 'Smooth' plumage (only obscure streaking on crown and back) with pale clear yellow underparts of adult female and imm unlike most other buntings, but in some cases may be indistinguishable from Red-headed Bunting (also rare) to which it is closely related. The most distinctive adult female Black-headed will show grey hood and rufous (vs. yellow) rump. The most obscure juv may appear more tawny-brown (Black-headed) vs. dull brown/greyish/greenish (Red-headed). **Voice** Song is a rather short, accelerating mixture of harsh rasping and sharp, clear notes with abrupt 'wrap-up', e.g. *zrit-zrit-zrit-chri-cheu-cheu-reh*. Call is a short, sharp *tsit*, *chuh*, or *chupp*. **Habitat** Nests in dry, open country with bushes and scattered trees, wooded edges and agricultural land; uses similar habitats on migration. **Behaviour** Similar to other buntings. **Status** Vagrant. A single sight record from Nömrög River valley, Dornod province, June 1994.

Red-headed Bunting *Emberiza bruniceps*   15–16 cm

ID In full frontal view, breeding male could be mistaken for Chestnut Bunting, but the latter has all-chestnut, unstreaked upperparts, while Red-headed has streaked olive back, white wing-bars and clear yellow rump. Adult female and imm Red-headed are very similar (in some cases indistinguishable) from closely related Black-headed Bunting, which see for best field marks. **Voice** Song and calls are essentially identical to Black-headed Bunting. **Habitat** Nests in dry, open habitats, including arid steppe and desert with scattered brush. **Behaviour** Generally similar to other buntings. Places nest in dense shrubs usually 10–50 cm above ground. Reported as not as shy as other buntings. **Status** Very rare and local breeding visitor to arid south-west, in Bulgan and Uyench River valleys, Khovd province and very rare passage migrant through breeding areas, presumably late April to early September.

Lapland Longspur *Calcarius lapponicus*   14–16 cm

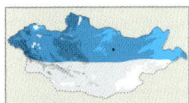

ID Breeding adult might be confused with adult Rustic Bunting, but head pattern is distinctive in both sexes, and longspurs never have rufous breast and flank streaking. In non-breeding birds, combination of reddish-brown face with black cheek-line, chestnut panel between two white wing-bars, and yellow bill with black tip is diagnostic. Note that majority of other bunting species are not present in Mongolia from late autumn to early spring when this species is present. **Voice** Song (unlikely to be heard in Mongolia) is a short, jangling warble. Call is a rather melodious *tee-ee*; also a rattled *prrrt*. **Habitat** Nests in Arctic tundra. In Mongolia, frequents dry, open habitats in high mountains, forest edges, grassy steppe and desert and cultivated fields where seeds are abundant. **Behaviour** Gregarious seed-eaters. Flocks of 20–2,000 wander in search of areas with thin snow cover and plentiful seeds. Often flocks together with Snow Buntings. **Status** Common winter visitor throughout country, with numbers lowest in south, mid-October to mid-April (mainly November to March). [Alt: Lapland Bunting]

Snow Bunting *Plectrophenax nivalis*   16–18 cm

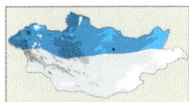

ID The combination of all-white head and black back is diagnostic in breeding male. Wintering birds can only be confused with White-winged Snowfinch, which also shows extensive white-and-black wing-tips in flight. However, latter species always has grey head and back, very unlike Snow Bunting's white head with rufous patches and rufous-and-black streaked upperparts. **Voice** The song is a twittering, repeated *hudidi feet feet feewh*. Call is similar to Lapland Bunting, but thinner: *prrrp*, or *tiriririt*. **Habitat** Nests in high Arctic tundra. Mongolian wintering birds prefer dry, open habitats, e.g. rocky mountain slopes, forest edges and grassy or brushy steppe. **Behaviour** Flocks of up to 300 seek out areas with thin snow cover where they can find an abundance of seeds. Often flock together with Lapland Longspurs and other granivorous birds. **Status** Uncommon to rare passage migrant and winter visitor to Northern Uvs and Great Lakes Depression east to Mongol Daguur Steppe; may also occur in the far east. Present mid-November to late March.

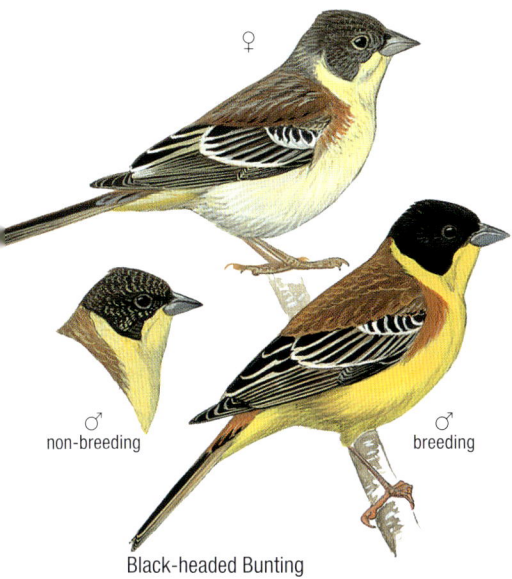

♀

♂
non-breeding

♂
breeding

Black-headed Bunting

♀

♂
non-breeding

♂
breeding

Red-headed Bunting

1st-winter

♂
non-breeding

♂
breeding

Lapland Longspur

♀
breeding

♂
non-breeding

♀
1st-winter

♂
breeding

Snow Bunting

RECENT VAGRANTS AND HYPOTHETICAL RECORDS

The species listed below have been recorded in Mongolia only recently. The details of these records have all been thoroughly reviewed and accepted by the Mongolian Bird Taxonomy and Rarity Committee.

Striated Heron *Butorides striata*

ID Small, stocky heron with short greenish-yellow legs, black cap, dark grey upperparts with greenish tinge and greyish underparts tinged buff. Similar to Black-crowned Night Heron, but separated by smaller size, scaly upperwing, loose dark nape-plumes and yellow lores. Juv is grey-brown with heavy streaks and spots above and below. **Status** Three records: Buir Lake, Dornod province (undated), Ongi River valley, Övörkhangai province, 26 May 2012, and Oyu Tolgoi mining area, Ömnögobi province, May 2013.

Intermediate Egret *Ardea intermedia*

ID Intermediate in size between Little and Great White Egrets. Most easily confused with Great White, but bill shorter and appearing stouter; at close range gape does not extend behind eye as in Great White. **Status** One reported from 3 km south of Baidrag River delta of Lake Bööntsagaan, Bayankhongor province, 8 June 2007.

Ruddy-breasted Crake *Porzana fusca*

ID Dull brown upperparts and ruddy face and breast, dark belly and undertail-coverts with white barring, and reddish-brown iris. Bill dark. Legs reddish-brown. Differs from Band-bellied Crake by less distinct white barring on lower belly and undertail-coverts, and strong bill. Juv is duller than adult with blue-tinged grey underparts and less barring on belly and undertail-coverts. **Status** Adult photographed between Luus and Khuld, Dundgobi province, 26 August 2016.

Band-bellied Crake *Porzana paykullii*

ID Adult differs from Ruddy-breasted Crake by paler reddish-brown legs and distinct bold barring on lower belly and undertail-coverts. Juv is duller brown with pale bill. **Voice** Drumming call *tototototo* at dusk and night. **Status** Adult male observed at a small pond 15 km north of Khalkh River, Dornod province, 22–24 May 2011. **Conservation** Considered Near Threatened globally.

Collared Pratincole *Glareola pratincola*

ID Very similar to Oriental Pratincole, but differs in showing contrast between darker outer wing and paler inner wing, prominent white trailing edge to secondaries in all plumages, and longer tail, reaching wingtips at rest. Juv resembles non-breeding adult, but has scaly pattern on upperparts and breast. **Status** A single bird seen on north-east shore of Bööntsagaan Lake, Bayankhongor province, 24 June 2016.

Buff-breasted Sandpiper *Calidris subruficollis*

ID General appearance is similar to juv Ruff, but significantly smaller and with shorter legs, more delicate build and shorter, finer bill. Juv Sharp-tailed Sandpiper is rufous above and on breast (vs. tawny-buff), and has distinct white supercilium and eye-ring. Buff-breasted Sandpiper typically forages in short dry grasslands rather than on lake shores or other wetlands. **Status** Adult photographed at Sum Höh Burd, Dundgobi province, 26 May 2014. **Conservation** Considered Near Threatened globally. **Taxonomy** Formerly placed in genus *Tryngites*.

Black-tailed Gull *Larus crassirostris*

ID Breeding adult has medium-grey upperparts, white rump, broad black subterminal band on white tail, pale iris, and four white spots on tips of folded wings. Non-breeding adult has dark streaks on crown and nape, and yellow bill with red-and-black ring. First-winter has brown-washed plumage with pale face, pinkish bill with black tip, pinkish legs, and broad black terminal band on tail. **Status** One at Döröö Lake, Dornod province, in April (year not known).

Red Turtle Dove *Streptopelia tranquebarica*

ID Adult male is unmistakable pinkish-rufous with contrasting blue-grey head. Female and juv are duller brown, but smaller, darker and shorter-tailed than Eurasian Collared Dove with dark (vs. pink) legs. Compare also Oriental Turtle Dove. **Status** One reported with flock of four Eurasian Collared Doves at Juulchin Gobi tourist resort, Ömnögobi province, 12–13 June 2007; another near Khongor sand dunes, Ömnögobi province, 5 July 2015.

European Roller *Coracias garrulus*

ID Unmistakable brilliant blue bird with contrasting rufous back, resembling small crow in general form. Perches on branches and wires and pounces shrike-like on insects and other prey. **Status** Adult identified at Zuun Khaatsain Khar Uzuur (800 m east of Yarantai military checkpoint), Khovd province, 18 May 2015.

Blue-cheeked Bee-eater *Merops persicus*

ID Resembles European Bee-eater in form and behaviour, but readily distinguished in all plumages by overall green plumage above (vs. chestnut and gold in European Bee-eater). **Status** Two adults photographed west of Ehiin gol research station, Bayankhongor province, 4 July 2014. **Conservation** Considered Near Threatened globally.

Ashy Minivet *Pericrocotus divaricatus*

ID Unmistakable forest species in Mongolia. Adult male has ashy-grey upperparts with black crown, nape, ear-coverts and eye-stripe. Forehead, underparts and undertail-coverts white variably washed light grey. Bill and legs black. Female has grey head including forehead, lacking black except on the lores. **Status** A single individual in poplar trees at Undain Gol, Ömnögobi province, 30 June 2016.

Ashy Drongo *Dicrurus leucophaeus*

ID Similar in form to Black and Hair-crested Drongos but readily distinguished by overall dark ashy-grey (vs. black) plumage. **Status** Immature of subspecies *D. l. leucogenis* identified in riparian forest in Terelj River valley, Töv province, 10 September 2013.

Hair-crested Drongo *Dicrurus hottentottus*

ID Distinguished from Black Drongo by heavier build and bill; hair-like feathers on crown; and in-curved outer feathers of forked tail. **Status** Adult photographed in poplar trees in valley of Ikh Nart, Dornogobi province, 7 June 2011; and adult *c.* 12 km west of Khalkh River, Dornod province, 4 June 2014.

Red-billed Starling *Spodiopsar sericeus*

ID Combination of grey body with contrasting white head and red legs and bill make adult male unmistakable. Adult female and juv are duller grey and lack white head but retain coloured bill base. Somewhat similar Daurian Starling has black bill and prominent white wing-bars. **Status** One photographed with flock of 22 White-cheeked Starlings in Uulbayan, Sukhbaatar province, 23 May 2012. **Taxonomy** Formerly placed in genus *Sturnus*.

Alpine Chough *Pyrrhocorax graculus*

ID Distinguished from adult Red-billed Chough by shorter yellow (vs. red) bill; however, juv Red-billed Chough has yellowish bill. In breeding season, typically occurs at higher altitude than Red-billed Chough. **Status** Eight birds photographed on high slopes of Indert mountain, Bayan-Ölgii province, 23 February 2014. The presence of these birds in appropriate habitat may indicate a previously unrecorded breeding population. [Alt: Yellow-billed Chough]

Eastern Crowned Leaf Warbler *Phylloscopus coronatus*

ID Distinguished from other leaf warblers with pale central crown-stripes by relatively large size, heavier bill and comparatively sluggish movements; from similar-sized Arctic Warbler by crown-stripe and very long supercilium, extending from above bill to nape. **Status** A singing male recorded from 15 km north of Khalkh River, Dornod province, 23 May 2011.

Chestnut-flanked White-eye *Zosterops erythropleurus*

ID Resembles *Phylloscopus* warbler, but combination of yellow throat, clear white underparts with bright chestnut flanks and heavy white eye-ring is diagnostic. **Status** Adult observed in elm trees at Ikh Nart, Dornogobi province, 19 September 2012.

White-capped Redstart *Phoenicurus leucocephalus*

ID A large, striking redstart, strongly associated with fast-flowing, high-altitude rivers and streams, descending to lower altitudes in non-breeding season. In all plumages only likely to be confused with adult Güldenstädt's Redstart, but lacks that species' prominent white wing-patch and has broad black terminal tail-band. **Status** One observed in Gegeet valley, at western end of Baruunsaikhan mountain, Ömnögobi province, 23 May 2009. **Taxonomy** Formerly placed in genus *Chaimarrornis*.

Slaty-backed Flycatcher *Ficedula hodgsonii*

ID Adult male is virtually unmistakable, with dark blue upperparts and rufous-orange underparts from throat to vent; also shows white patches at base of black tail. Somewhat similar male Black Redstart has black throat and bright rufous tail. Female and juv of this and other vagrant *Ficedula* flycatchers are notably nondescript and should be identified with extreme caution. **Status** Adult male reported in large plantation near town of Dalanzadgad, Ömnögobi province, 22 June 2015.

Rosy Pipit *Anthus roseatus*

ID Breeding adult readily distinguished by dull rosy breast; non-breeding birds share bold back and breast streaking only with Red-throated and Pechora Pipits, but have distinctive olive tone to upperparts and comparatively prominent white supercilium. **Status** One reported near an artificial pond at Oyu Tolgoi, Ömnögobi province, 6 May 2012.

REFERENCES

Ayé, R., Schweizer, M. & Roth, T. (2012). *Birds of Central Asia: Kazakhstan, Turkmenistan, Uzbekistan, Kyrgyzstan, Tajikistan, Afghanistan.* Bloomsbury, London.

Bold, A. (1973). Birds of Mongolia. *Scientific Proceedings of the Institute of Biology of the Mongolian Academy of Sciences* 7: 139–167. [in Mongolian]

Bold, A. & Mainjargal, G. (2006). Very rare birds in Mongolian territory. *Scientific proceedings of the Institute of Biology of the Mongolian Academy of Sciences* 26: 78–81. [in Mongolian]

Bold, A. & Stepanyan, L. S. (1988). A Checklist of Birds of the Mongolian People's Republic. *Ornithology* 5: 23. [in Russian]

Bold, A., Boldbaatar, Sh. & Tseveenmyadag, N. (2002). A checklist of birds of Mongolia. *Birds, Amphibians and Reptiles of Mongolia* 1: 19–38. [in Mongolian]

Bold, A., Sumiya, D. & Tseveenmyadag, N. (1983). Additional changes to the birds of Mongolian People's Republic. *Scientific proceedings of the Institute of the General and Experimental Biological Institute* 15: 79–88. [in Mongolian]

Bold, A., Tseveenmyadag, N., Boldbaatar, Sh. & Mainjargal, G. (2007). Checklist of Mongolian birds in ten different languages. *State Terminological Proceedings* 2: 150. [in Mongolian]

Bold, A., Tseveenmyadag, N., Boldbaatar, Sh., Sumiya, D., Gombobaatar, S., & Mainjargal, G. (2001). Review of names of species, genera, family and orders of birds of Mongolia. *Proceedings of the Institute of Philology* 1: 80–91. [in Mongolian]

Boldbaatar, Sh. (2000). Very rare and rare bird surveys in river valleys of the Eg Tavagatai tributary and Khangai mountains. *Central Asian Ecosystems* 1: 179–180. [in Mongolian]

Boldbaatar, Sh. (2001). New records of birds in Mongolia in the last decade. *Biodiversity of the Mongolian Plateau and Adjacent Territory.* Ulaanbaatar. [in Mongolian]

Boldbaatar, Sh. (2002). New records for the Mongolian avifauna. *Birds, Amphibians and Reptiles of Mongolia* 1: 45–49. [in Mongolian]

Boldbaatar, Sh., Gombobaatar, S., Mainjargal, G. & Uuganbayar, Ch. (2013 and 2016). *Mongolian Red Book.* Ministry of Environment and Tourism, Ulaanbaatar. [in Mongolian and English]

Brazil, M. (2009). *Birds of East Asia.* Christopher Helm, London.

Brazil, M. (2018). *Birds of Japan.* Helm, London.

Cramp, S. & Simmons, K. E. L. (1980). *Handbook of the Birds of Europe, the Middle East and North Africa* (Vol. 2). Oxford University Press, Oxford.

Dementiev, G. P. & Gladkov, N. A. (1951). *Birds of the Soviet Union* (Vol. 1). Nauka, Moscow. [in Russian]

Fomin, V. E. & Bold, A. (1991). *Catalogue of the Birds of the Mongolian People's Republic.* Nauka, Moscow. [in Russian]

Forsman, D. (1999). *The Raptors of Europe and the Middle East.* T & A D Poyser, London.

Gelang, M., Cibois, A., Pasquet, E., Olsson, U., Alström, P. & Ericson, P.G.P. (2009). Phylogeny of babblers (Aves, Passeriformes): major lineages, family limits and classification. *Zoologica Scripta* 38: 225–236.

Gombobaatar, S. (2010). *A Photographic Guide to the Birds of Mongolia.* Mongolian Ornithological Society, Ulaanbaatar.

Gombobaatar, S. & Amartuvshin, P. (2011). Birdlife of Mongolia. *Sustainable Tourism in Mongolia.* 2011 (special edition): 21-26.

Gombobaatar, S. & Bayanmunkh, D. (2016). *Birds Mongolia: an Annotated Checklist.* Mongolica Publishing and Mongolian Ornithological Society, Ulaanbaatar.

Gombobaatar, S. & Usukhjargal, D. (2015). *Birds of Hustai National Park, Mongolia.* 2nd edition. Hustai National Park and Mongolian Ornithological Society, Ulaanbaatar.

Gombobaatar, S., Boldbaatar, Sh. Tseveenmaydag, N. Bayarkhuu, S. & Usukhjargal, D. (2014). Mongolian bird rarities in 2013-2014. *Ornis Mongolica* 3(432): 15-25.

Gombobaatar, S. (compiler), Brown, H. J., Sumiya, D., Tseveenmyadag, N., Boldbaatar, Sh., Baillie, J. E. M., Batbayar, G., Monks, E. M., Stubbe, M. (eds) (2011). *Summary Conservation Action Plans for Mongolian Birds.* Regional Red List Series Vol. 8. Zoological Society of London, Mongolian Ornithological Society and National University of Mongolia, Ulaanbaatar.

Gombobaatar, S. & Monks, E. M. (compilers), Seidler, R., Sumiya, D., Tseveenmyadag, N., Bayarkhuu, S., Baillie, J. E. M., Boldbaatar, Sh., Uuganbayar, Ch. (eds). (2011). *Regional Red List Series Vol. 7.* Zoological Society of London, National University of Mongolia and Mongolian Ornithological Society, Ulaanbaatar.

Grimmett, R., Inskipp, C. & Inskipp, T. (2011). *Birds of the Indian Subcontinent.* Christopher Helm, London.

Kishinskii, A. A., Fomin, V. E., Bold, A. & Tseveenmyadag, N. (1982). Birds of the Munkhkhairkhan mountain massif (Mongolian People's Republic). *Zoological Surveys in the Mongolian People's Republic.* Nauka, Moscow. [in Russian]

Kozlova, E. V. (1930). *Birds of South-Eastern Zabaikaliya, Northern Mongolia and Central Asia.* USSR Academy of Sciences, Leningrad. [in Russian]

Kozlova, E. V. (1932). Birds of Khangai high mountains. *Proceedings of the Mongolian Commission.* No 3. USSR Academy of Sciences, Leningrad. [in Russian]

Kozlova, E. V. (1975). *Birds of Zonal Steppes and Deserts of Central Asia.* Nauka, Leningrad. [in Russian]

MacKinnon, J. & Phillips, K. (2000). *A Field Guide to the Birds of China.* Oxford University Press, Oxford.

Madge, S. & Burn, H. (1988). *Wildfowl: An Identification Guide to the Ducks, Geese and Swans of the World.* Christopher Helm, London.

Nyambayar, B. & Tseveenmyadag, N. (eds) (2009). *Directory of Important Bird Areas in Mongolia: Key Sites for Conservation.* Admon Publishing, Ulaanbaatar. [in Mongolian and English]

Olsen, K. M. & Larsson, H. (2003). *Gulls of Europe, Asia and North America.* Christopher Helm, London.

Ostapenko, V. A., Gavrilov, V. M., Bold, A. & Tseveenmyadag, N. (1979). Distribution and biology of several species of reed warblers in western Mongolia. *Ornithology* 14: 195–196. [in Russian]

Piechocki, R., Stubbe, M., Ulenhaut, K. & Sumiya, D. (1982). Contribution to the Avifauna of Mongolia. *Mitteilungen aus dem Zoologischen Museum Berlin Supp. 58. Ann. Orn.* 6: 3–53. [in German]

Porter, R. F., Christensen, S. & Schiermacker-Hansen, P. (1996). *Field Guide to the Birds of the Middle East.* T & A D Poyser, London.

Potapov, R. L. (1986). The ornithofauna in the Mongolian Altai and surrounding territories. Distribution and biology of birds of the Altai and Far East. *Proceedings of the Zoological Institute, Russian Academy of Sciences* 150: 57–73. [in Russian]

Przewalskii, N. M. (1876). *Mongolia and Country of Tangut. A three-year expedition in eastern mountainous Asia.* V. II. Sanktpterburg. Russia. [in Russian]

Purevsuren, Ts., Dashnyam, B., Dorjderem, S., Bataa, D. & Amarsaikhan, S. (2013). Rare birds of Oyu Tolgoi project area. *Ornis Mongolica* 2: 47-56.

Robson, C. (2009). *A Guide to the Birds of Southeast Asia.* Princeton University Press, Princeton.

Samiya, R. (2013). Documented new records on some bird species in Mongolia. *Ornis Mongolica* 2: 57-59.

Shagdarsuren, O. (1983). *Raptors of Mongolia.* Ulaanbaatar. [in Mongolian]

Shirihai, H. Gargallo, G. & Helbig, A. (2001). *Sylvia Warblers.* Christopher Helm, London.

Smirinskii, S. M., Sumiya, D. & Boldbaatar, Ts. (1991). Ornithological observations in Eastern Mongolia. *Ornithology* 25: 116–126. [in Russian]

Snow, D.W. & Perrins, C.M. (1998). *The Birds of the Western Palearctic: Non–passerines.* Vol. 1. Oxford University Press, Oxford.

Stenzel, T., Stubbe, M., Samjaa, R. & Gombobaatar, S. (2005). Quantitative investigations on bird communities in different habitats in the Orkhon-Selenge valley in Northern Mongolia. *Erforschung Biologischer Resourcen der Mongolischen Volksrepublik* 9: 311–391. [in German]

Stepanyan, L. S. (2003). *Conspectus of Ornithological Fauna of Russia and Neighbouring Territories.* Nauka, Moscow. [in Russian]

Stephan, B. (1994). Ornithological studies in Mongolia. *Mitteilungen aus dem Zoologischen Museum Berlin* 70, *Suppl. Ann. Orn.*18: 53–100. [in German]

Sumiya, D. (2002). Birds of Huvsgul Lake and its surroundings: non-passerines. (Non-Passeriformes, Aves). *Birds, Amphibians and Reptiles of Mongolia* 1: 126–178. [in Mongolian]

Sushkin, P. P. (1938). *Birds of Russian Altai and Adjacent Territories of the North-Western Part of Mongolia.* Nauka, Moscow. [in Russian]

Svensson, L., Mullarney, K. & Zetterström, D. (2009). *Collins Bird Guide* 2nd edition. HarperCollins, London.

Tseveenmyadag, N. & Bold, A. (2006). Additional changes to the species list of birds in Mongolia. *Scientific Proceedings of the Institute of Biology of the Mongolian Academy of Sciences* 26: 129–133. [in Mongolian]

Tseveenmyadag, N., Bold, A., Boldbaatar, Sh. & Mainjargal, G. (2005). *Field Guide to the Birds of Hentii Mountain Range.* Mongolian Academy of Sciences, Ulaanbaatar. [in Mongolian]

Tseveenmyadag, N., Bold, A., Fomin, V. E. & Ostapenko, V. A. (2000). Birds of the Onon, Ulz and Khalkh river basins. *Scientific Proceedings of the Institute of Biology of the Mongolian Academy of Sciences* 22: 153–160. [in Mongolian]

Tugarinov, A. Ya. (1932). Birds of eastern Mongolia from observations on the expedition of 1928. *Proceedings of the Mongolian Committee* 1: 1-46. [in Russian]

Vaurie, C. (1965). *The Birds of the Palearctic Fauna: Vol. 2, Non-passeriformes.* H.F. & G. Witherby, London.

Voous, K. H. (1977). *List of Recent Holarctic Bird Species.* British Ornithologists' Union, Peterborough.

Yamazaki, T., Nitani Y., Murate, T., Lim, K. C., Kasorndorkbua, C, Rakhman, Z. & Gombobaatar, S. (2012). *Field Guide to Raptors of Asia: Migratory Raptors of Oriental Asia.* Vol. 1. Asian Raptor Research and Conservation Network, Japan.

INDEX

S